T0135597

Analysis of process-induced distortions and residual stresses of composite structures

Dem Fachbereich Produktionstechnik

der

UNIVERSITÄT BREMEN

zur Erlangung des Grades
Doktor-Ingenieur
genehmigte

Dissertation

von

Christian Brauner (M.Sc.)

Gutachter: Univ.-Prof. Dr.-Ing. Axel S. Herrmann, Universität Bremen

Univ.-Prof. Dr.-Ing. habil. Lothar Kroll, Technische Universität Chemnitz

Tag der mündlichen Prüfung: 26. August 2013

Science-Report aus dem Faserinstitut Bremen

Hrsg.: Prof. Dr.-Ing. Axel S. Herrmann

ISSN 1611-3861

Bibliografische Information der Deutschen Nationalbibliothek

Die Deutsche Nationalbibliothek verzeichnet diese Publikation in der
Deutschen Nationalbibliografie; detaillierte bibliografische Daten sind
im Internet über http://dnb.d-nb.de abrufbar.

ISBN 978-3-8325-3528-5

Logos Verlag Berlin GmbH
Comeniushof, Gubener Str. 47,
10243 Berlin
Tel.: +49 030 42 85 10 90
Fax: +49 030 42 85 10 92
INTERNET: http://www.logos-verlag.de

Danksagung

Die vorliegende Arbeit entstand in der Zeit vom August 2009 bis zum November 2012 in der Anstellung als Wissenschaftlicher Mitarbeiter am Faserinstitut Bremen e.v. Diese Zeit am Faserinstitut habe ich als sehr anregend empfunden, besonders durch den kollegialen Umgang, die Möglichkeiten der interdisziplinären Zusammenarbeit, die exzellente Ausstattung und den Freiraum, welcher mir durch Professor Dr.-Ing. Axel S. Herrmann bei dem Erstellen der Dissertation eingeräumt wurde.

Mein besonderer Dank gilt deshalb Professor Dr.-Ing. Axel S. Herrmann für die Möglichkeit dieser Promotion und die hilfreiche Unterstützung. Desweiteren möchte ich mich bedanken bei Univ.-Prof. Dr.-Ing. habil. Lothar Kroll für die Übernahme des Koreferates. Ich danke Herrn Prof. Dr.-Ing. Thomas Hochrainer für die Übernahme des Prüfungsvorsitz sowie Herrn Dr.Ing. Jörg Jendrny, Frau Dipl.Ing. Johanne Hesselbach und Herrn cand. Ing. Timo Fischer für ihr Engagement in der Prüfungskommission.

Ein herzlicher Dank geht an Dipl-Ing. Christoph Hoffmeister; welchen ich für die vielen wertvollen Diskussionen und die fachliche Unterstützung sehr schätze. Es fällt mir schwer, einzelne Mitarbeiter und Freunde aus dem Faserinstitut aufzuzählen, daher möchte ich allen für die Hilfen, Anregungen und die Ablenkungen Danke sagen. Ohne Euch hätte ich die vergangenen 3 Jahre nicht genossen.

Dr.-Ing. Stefan Bauer möchte ich speziell für die Möglichkeiten der Anwendung der entwickelten Methoden im industriellen Umfeld danken. Ich schätze die Diskussion, welche wir im Team CMF/Landklappe geführt haben sehr.

Mein allergrößter Dank geht an meine Eltern, meinen Bruder Florian und meiner Frau Steffi, die mich stets bestärkt haben, wenn ich an mir gezweifelt habe und ohne deren Unterstützung diese Arbeit nicht möglich gewesen wäre.

Gewidmet meiner Mutter

Abstract

The increased application of composite materials in lightweight structures leads to new integral design of structural parts, using the example of, an integral composite landing flap of an Airbus A320 aircraft. This offers possibilities to simplify the process chain, to decrease manufacturing costs and to have fibre fair structural design. The critical disadvantage of large integral designs is that process-induced deformations are a risk factor during the design phase of the manufacturing process, because rework is not always possible. Especially, the aircraft industry with its demands on high qualities / tolerances and the application of hot curing resin systems, requires knowledge-based methods for virtual process design to avoid time- and effort-consuming iterations.

This thesis contributes to the understanding of the mechanism behind process-induced distortions and stresses related to the Resin Transfer Moulding (RTM) manufacturing process of an integral composite landing flap. The aim is to comprehend the phenomena, to identify related parameters and to present compensation strategies. Therefore, the first part of this thesis is an experimental study about the behaviour of material properties during the manufacturing process of the single components and the composite. It concludes with constitutive equations for single process parameters and their associated homogenisation approach for the composite properties. During the manufacturing process, engineering constants of the matrix are changing and influenced by a high number of effects. In the second part of this thesis, there is a detailed study on the micro-scale in order to understand the discovered effects and their impact on the homogenised macro-scale. In the third part of this thesis a simulation strategy and a viscoelastic material model for the macro-scale have been derived. This developed material model integrates a dependency of the time - temperature - polymerisation and fibre volume content on the relaxation behaviour of residual stresses in a transversal isotropic reinforced material. The model is validated in two test cases at the coupon level and limitations of use are discussed.

In the fourth part of the thesis the tested virtual model is used to analyse process-induced distortions and stresses of an integral composite landing flap and demonstrates the application in an industrial environment. In the fifth part the deterministic model is used in connection with probabilistic methods, such as virtual sensitivity analysis, variability analysis and process optimisation, to demonstrate further application of the developed approach.

On the one hand, the specific advantage of the developed model is a transient analysis of the process with an integration of relevant factors to achieve reliable results for process-induced distortions and stresses. On the other hand, homogenised engineering properties are as a result available. Therefore, the method contributes to a virtual process design and can also be the first step of more accurate structural design.

Zusammenfassung

Der verstärkte Einsatz von Faserverbundwerkstoffen im Leichtbau führt zu neuen integralen Bauweisen von Bauteilen, wie der integral gefertigten Faserverbundlandeklappe. Hierbei wurde die bestehende differentiell gefertigte Landeklappe des Airbus A320 durch eine integrale Bauweise, welche in einem einzigen Fertigungsschritt gefertigt wird, ersetzt. Dieses eröffnet Möglichkeiten, bestehende Prozessketten zu vereinfachen, Fertigungskosten zu senken, sowie fasergerechte Konstruktionen auszuführen, welche in ihrem Strukturverhalten das anisotrope Verhalten des Werkstoffes optimaler ausnutzen. Ein entscheidender Nachteil großer integraler Bauteile ist, dass auftretende, prozessbedingte Verformungen einen erheblichen Risikofaktor bei der Entwicklung des Herstellungsprozess darstellen, da eine nachträgliche Korrektur der äußeren Form des Bauteils nicht möglich ist. Speziell in der Luftfahrtindustrie mit den hohen Qualitätsanforderungen und den warmaushärtenden Matrixsystemen ist ein Bedarf an wissensbasierten Auslegungsmethoden vorhanden, welche das Prozessverständnis erhöhen, den Fertigungsprozess virtuell abbilden und somit zeitaufwändige und teure Iterationsschleifen bei der Entwicklung von Fertigungsprozessen vermeiden.

Diese Dissertation liefert einen Beitrag, den Herstellungsprozess Resin Transfer Moulding (RTM) am Beispiel einer integral gefertigten Faserverbundlandeklappe zu verstehen, verzugsrelevante Parameter zu identifizieren und Strategien zur Kompensation vorzuschlagen. Der erste Teil der Arbeit ist eine experimentelle Studie des Materialverhaltens im Fertigungsprozess und die Ableitung von konstitutiven Beschreibungen. Diese wird sowohl für die einzelnen Komponenten des Verbundes betrieben, wie auch für den Verbund mit den nötigen Homogenisierungsmethoden. Da sich im Herstellungsprozess die Eigenschaften der Einzelkomponente Matrix erst entwickeln, werden die gemessenen Effekte auf der Mikroskala diskutiert und, im zweiten Teil der Arbeit, ihr Einfluss auf die homogenisierten Eigenschaften auf der Makroskala dargestellt. Im dritten Teil der Arbeit wird eine Simulationsstrategie entwickelt und ein viskoelastisches Materialmodell abgeleitet. Dieses Materialmodell beschreibt das Relaxationsverhalten von Eigenspannungen, basierend auf den Einflüssen Zeit, Temperatur, Polymerisationsgrad und Faservolumengehalt. Die Einsatzgrenzen des Modells werden anhand von Validierungstests aus der Couponebene diskutiert. Dieses virtuelle Modell wird im vierten Teil der Arbeit zur Analyse prozessbedingter Verformungen und Eigenspannungen einer integral gefertigten Faserverbundlandeklappe genutzt und demonstriert die Anwendung des Modells im industriellen Umfeld. Im fünften und letzten Teil der Arbeit wird ein Ausblick gegeben, die entwickelte deterministische Methode in Verbindung mit probabilistischen Ansätzen zur virtuellen Sensitivitätsanalyse, Robustheitsanalyse und Prozessoptimierung zu nutzen.

Die Besonderheiten der entwickelten Ansätze liegen darin, dass diese durch die transiente Darstellung des Prozesses in der Lage sind, Inhomogenität im Bauteil zu berücksichtigen und, dass als Ergebnisse nicht nur prozessbedingte Verformungen und Eigenspannungen darstellbar sind, sondern auch resultierende Materialeigenschaften. Hiermit liefert die entwickelte Methode einen Beitrag, virtuell Fertigungsprozesse darzustellen und genauere Eingangsgrößen für die Strukturanalyse zu liefern.

1   Introduction

High performance carbon fibre reinforced composite materials are widely used in aerospace structures due to their advantages compared to conventional, metallic, materials: they are lightweight, have high stiffness / strength, and are very durable. In order to attain the full potential of these materials, besides the given advantages, the disadvantages of relatively high material and manufacturing costs must be compared. Monolithic designs tend to become largely integral in order to achieve lower manufacturing costs [1]. Consequently, reducing manufacturing costs is possible, due to a lower overall part count and simplified designs, e.g. reducing the number of joints and fasteners significantly. For highly integral monolithic structures a major challenge is the development of a robust manufacturing process to produce high quality structures. An integral structure must be adapted to the tolerance requirements, because there is no possibility to change the final geometry in the assembling step, afterwards. Process-induced distortions are great risk factors for these types of structures related to tolerance requirements, manufacturing costs and process time. In the current development of manufacturing processes, a high effort is raised to find the optimum of process parameters to achieve the required tolerances and quality. Most of the time, a trial-and-error method is used, which is sometimes a random process. The development process starts, usually, with a parameter set which is based on the manufacturer and material provider specifications. Different variations are then made with the process parameters until an acceptable quality is reached or time / funding are consumed. At this particular point the best parameter sets are chosen. Therefore, very conservative process parameters are selected and the tolerances of the parameters are very high to ensure a repeatable process with qualities inside the so-called "process window". The main advantage of such a trial-and-error method is that it does not require a high level of knowledge about the process itself. The disadvantage is that it is not practical to perform all possible variations in order to find the real optimum of the process, because a lot of mechanical and physical testing is needed to determine and evaluate the product quality. As a result it is difficult to transfer the process knowledge to new materials, processes or parts. Also, there is no information about the disturbance of the process in view of the variability of material or process parameters.

Manufacturing process simulation is an essential part of product development to increase the knowledge of the physical effects during the process and to make them virtual evident. Consequently, a virtual process chain can enhance quality of components, decrease process development time / costs and lead to more efficient process designs with reduced cycle times and scrap rates.

Finally, process simulation is the first part of the full virtual design which should provide information not only about part distortions and residual stresses, but also about resulting material properties and their inhomogeneous distribution inside of a component. Therefore, the process simulation could also increase the reliability and accuracy of the structural design.

1.1 Objective

The objective of this thesis is to develop a virtual process chain for the Resin Transfer Moulding process to represent all physical effects which lead to process-induced deformations and stresses. The requirements of this integrated model are that it should incorporate all sources of physical effects which lead to process-induced deformations and residual stresses. It should be applicable for use in the industrial environment as well as for large integral structures. Additionally, the model shall provide the following benefits to production engineering:

- Reproducing the physical behaviour to represent process-induced deformations and stresses with focus on the viscoelastic relaxation effect.
- The characterisation of necessary material parameters should be as easy as possible.
- Applicable to perform sensitivity, variability and optimisation studies.
- Providing a virtual characterisation on the ply level with incorporated local variations of material parameters depending on the process conditions. This enables to perform so called "As-Built" analysis. This possibility offers to re-qualify parts with failed process parameters or to start the structure design certification with more reliable material input data.

1.2 Structure of the work

This work is structured in the following way: in chapter 2, a general overview is given about the manufacturing processes, main sources and relevant factors for process-induced distortions and residual stresses, and analysis methods. Because of the complexity of this topic, the origin of deformations and stresses is focused on the manufacturing process Resin Transfer Moulding (RTM). Therefore, this process is explained in detail, whereas, other manufacturing processes are not considered.

Chapter 3 and 4 will focus on the characterisation of the used materials in the RTM process. Accordingly, different experimental studies are performed for pure resin and for composite. In derivation of the experiments, necessary constitutive equations and parameters are evaluated for the description of important factors, such as cure kinetics, glass transition temperature, chemical shrinkage, thermal conductivity, heat capacity, cure-dependent engineering properties, etc. Due to the nature of the RTM process a homogenisation of the fibre and matrix materials is performed during the process. Chapter 4 closes with necessary homogenisation methods which take into account changing resin properties during the process.

The following chapter 5 discusses different effects of the experimental characterisation on the formation of residual stresses on the micro- level. This detailed study was conducted to evaluate the sensitivity of different material parameters in order to interpret their physical effects and to verify homogenised residual stresses on the macro-level.

Further in chapter 6, the analysis method on the macro-level is presented and discussed in detail. The method is divided into analysis modules, thermodynamic and mechanical. A validation for both modules is conducted on the coupon level using two test cases. Firstly, a proof is performed on the correct determination resulting engineering properties and stresses using tensile tests on unidirectional and secondly, to validate process-induced deformations the distortion of an unsymmetrical laminate is compared and discussed.

In chapter 7, the validated method is applied to a test structure, namely to analyse the manufacturing process of an integral RTM Composite Multispar high-lift Flap (CMF). The test structure has the length of 7.5 m. The application of the analysis method in the industrial environment on this particular box structure presents the possibilities to perform parameter studies on process parameters to reduce process time, to evaluate risk parameters and to study their influence to improve quality.

Chapter 8 gives a perspective on the use of the analysis method with respect to nondeterministic methods to evaluate the sensitivity of process parameters using Design of Experiment and the process variability using Monte Carlo Simulation. For the Monte Carlo Simulation the standard deviation of the parameters evaluated in chapter 3 are used to compute the impact on the variability. These studies are compared with the variability of the experimental validation test cases of chapter 6.

At last, the results are summarised and a prospect is given of necessary research to improve the reliability of the presented work in chapter 9.

2 State of the art

In this chapter, a compact overview of different manufacturing processes for composite materials is given. After this, related to the objective of this thesis, the main reason for process-induced deformations and stresses are explained and different approaches are presented to analyse them.

2.1 Manufacturing processes for composites

A variety of different processes is available to manufacture thermoset fibre reinforced composite parts. In general, there are process methods which use Prepeg materials (pre-impregnated fibre and high viscosity matrix systems) or dry textile materials which are impregnated by infusion (vacuum driven) or injection (pressure driven). Dry textile-based processes are called Liquid Composite Manufacturing (LCM) and can be classified into open mould concepts like Resin Transfer Infusion (RTI) and closed mould concepts like Resin Transfer Moulding (RTM). To reach high fibre volume contents, accelerated resin impregnation times and a minimum of voids, additional pressure is applied. This can easily be performed in the case of a closed mould; in the case of an open mould an autoclave is needed to add pressure on the vacuum bag. Beside this, the RTM process has the advantage that a heating system can be applied by a heated press which has lower energy consumptions than an autoclave. A further development of the RTM process is to integrate the heating system into the mould itself using, e.g. a fluid heating system. This is the so-called FAST-RTM process. Using this technology, uniform temperature can be applied with accelerated heat or cooling rates up to 30°C/min [3].

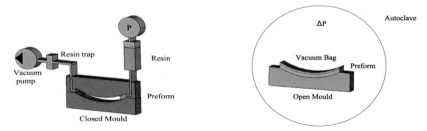

Figure 1 RTM process – RTI process

The RTM process is a suitable process for mass production with high quality and a minimum of manufacturing costs. In the following parts of this thesis, all developed simulation methods focus on the RTM process and on materials which are used for aerospace applications.

For the RTM process the following process steps have to be fulfilled. Dry textile preforms are placed into a mould, heated to an injection temperature of 100 to 120°C, injected, heated to the curing temperature of 180°C, cured for 2h and cooled down to room temperature. During the curing the thermoset resin passes through three different morphologic states (Figure 2). The resin converts from a liquid (I) to a rubbery state (II) and finally into a solid state (III). These changes are gradual, but from an engineering point of view the transition can be defined as the gel point (A) and the vitrification point (B). As a consequence of these phase transitions, a material model has to include all phase transitions.

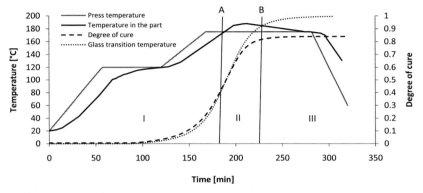

Figure 2 Curing process [4]

2.2 Main sources and factors for process-induced distortions and residual stresses

The distinctive feature of composite materials is that the resulting material properties can be designed by selection of different fibre / fabric and matrix materials. Thermoset polymers are used as matrix material and due to the high requirements of the aerospace industries, epoxy resin systems are preferred which are thermally stable. This leads to the point that, typically, resin systems are chosen with high curing temperatures around 180°C to reach glass transition temperatures above 120°C (for hot / wet conditions). These high curing temperatures are one of the main sources for process-induced distortions and stresses for aerospace application. The second main source is the chemical induced shrinkage of the resin during curing. Therefore, the main sources of process-induced distortions and stresses for thermoset composite materials are the thermal contraction of the composite, the cure shrinkage of the matrix material. They are influenced by several factors.

The factors can be manifold and different researchers have attempted to classify them into different categories such as first and second order [12], or intrinsic and extrinsic [86]. Additionally, several discussions can be found about the sensitivity of several factors e.g. Svanberg et al. [21].

Related to the layup or the part geometry, these factors / sources can lead to two different types of process-induced deformations. The first type is the so called spring-in effect, a change of the final value of an angle section. This is based on the geometry of the corner, fibre volume gradients over the thickness, thermal or cure gradients in thickness directions and tool / part interactions. The second type is classified as warpage effect, namely a distortion of initial flat part, based on unbalanced or unsymmetrical laminates. Therefore, all factors which can influence the symmetry conditions of a layup are critical. The following factors can be listed: preforming errors (angle errors, undulations, draping errors, gaps), thermal or cure gradients in plane of the part, tool / part interactions (if the tool is only on one side e.g. open mould concepts) and fibre volume content gradients over the thickness (different compaction by pressure gradients, resin flow) [21].

The effects of these factors / sources on process-induced stresses may become large enough to lead to fibre matrix debonding, matrix failure or delamination. In general, defects on the micro scale, such as matrix fibre debonding or small matrix cracks, are hardly measurable and, consequently, only noticeable in reduced stiffness or strength properties (Fig 3.).

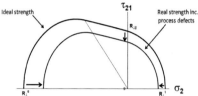

Figure 3 Influence of process defects on material strength [2]

The superposition of process-induced and mechanical stresses can also lead to critical situations. In composite materials, residual stresses can change the failure behaviour [6]. Using criteria like the Puck criterion, a complex stress state can be condensed and displayed by a vector of material effort. In general, this vector starts at the origin of the stress coordinate system and indicates failure if it touches the envelop failure curve. In the case there is residual stress, the vector of effort does not start at the origin anymore (Fig. 4). By neglecting process-dependent stresses, the effort value can change significantly, e.g. the mode of transverse failure can change from mode B to the more critical failure mode C, which can lead to interlaminar delamination failure.

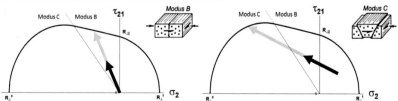

Figure 4 Influence of residual stress on the failure behaviour [2], left side - without residual stress, right side - change of the material effort due to residual stresses

As shown in the previous section, many different manufacturing methods can be used to process a composite part. In this thesis, an analysis method will be developed, with focus on the RTM process. Therefore, it is necessary to access the main sources and factors in relation to the RTM manufacturing process and the objective structure which should be analysed to identify the proceeding, the needed material characterisation and enhancement of material models. The objective structure is the composite multi spar flap which can be classified as a box structure. A box structure is, in general, sensitive to the warpage and spring-in effects. In the case of a composite flap the spring-in effect can be neglected, because no sharp corners exist on the outer shape. Consequently, warpage is the critical type of deformation. Many researchers have identified [86, 4, 5, 21, 42] the layup compaction, resin flow and tool part interaction as important factors influencing the symmetrical conditions of a layup which tends to warpage. In the given application, the layup compaction is performed during the preforming using a binder, which melts at a lower temperature of around 70°C during the process. At this moment layup compaction is then released in the cavity of the RTM mould. Therefore, effects from the compaction which create fibre volume content gradients in thickness directions can be neglected. Resin flow based on pressure gradients for autoclave processing parts is a critical effect which can lead to fibre volume content gradients in thickness direction. In the RTM process, resin flow is critical to ensure form filling, but during the curing process no resin flow occurs. Accordingly, fibre volume content gradients based on resin flow can be neglected.

The effect of tool-part interaction is also more critical in case of autoclave processed parts, because the tool is only on one side. This induces an unsymmetrical condition which leads to warpage. In the case of the RTM process, the tool part interaction is on both sides of the composite part. Additionally, to increase cost efficiency of RTM processes, parts are removed from the tool in hot conditions to reduce cycle time or to ensure release properties for box structures with integrated mandrels. Consequently, tool part interactions are not a main driver of process-induced deformation for RTM processed parts.

The most critical elements of a RTM process for a box structure are the non-uniform temperature conditions in the mould and composite part during the process, especially during the heating phase from injection temperature to curing temperature. If the applied heating rate will be too fast, a non-uniform curing influences the warpage. Therefore, the demand can be identified to compute the correct temperature during the process.

A second critical element is related to the nature of a box structure. A box structure is made with a mandrel inside the box. Largely, the fixation of such a mandrel is not simple and the condition of a fixed or known cavity is no longer fulfilled. In the case changes of thickness, the fibre volume content is influenced and, therefore, all composite properties such as Young's modulus, thermal contraction, chemical induced shrinkage, relaxation of stresses etc. can vary. Hence, a material model is needed which can be considered as a variegating fibre volume content, but also should incorporate all sources of physical effects dependent on temperature and curing reaction.

In summary, the analysis method for the RTM process of a box structure should be able to evaluate the right temperature during the process. Consequently, characterisation and modelling of relevant thermo-chemical parameters like curing reaction, heat of reaction, heat capacity, thermal conductivity and density is needed.

Furthermore, the method should be as flexible to take into account all effects which disturb the symmetrical conditions on the geometrical level, and also on the laminate level. Therefore characterisation and modelling is required for mechanical parameters like modulus, thermal contraction, chemical induced shrinkage, relaxation etc.

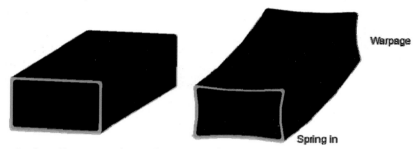

Figure 5 Deformed box structure due to a thinner upper side

2.3 Material models

In the following section, different material models for analysing process-induced deformations and residual stresses are presented. In general, the material models can be classified into three groups: elastic models, incrementally elastic models and viscoelastic models [16].

Elastic models
A fundamental source of shape distortion is the thermal expansion of the anisotropic or orthotropic material. For most composite resin fibre combinations the fibre has a small thermal expansion compared to the polymer matrix. Redford and Renick [6] have formulated a simple analytical formulation that can be applied to compute the spring-in angle $\Delta\theta$ of a curved section if other influences such as chemical shrinkage can be neglected

$$\Delta\theta = \theta \left[\frac{(\alpha_T - \alpha_R)\Delta T}{1 + \alpha_R \Delta T} \right].$$

(2.1)

In this approach α_T and α_R are the thermal expansion coefficients of the curved section in tangential and radial direction. It is assumed that the material properties are orthotropic and uniform throughout the thickness. The thermal expansion for a unidirectional non-crimp CFRP composite ply in fibre direction is around $-0.4 \cdot 10^{-6} \, 1/K$ and in transverse direction $30 \cdot 10^{-6} \, 1/K$. This implies that thermal expansion is to be a first order effect on the system. An enhancement of this approach was performed to include the effect of chemical shrinkage. The equation was extended by a term for the chemical shrinkage using ϕ_T and ϕ_R

$$\Delta\theta = \theta \left(\left[\frac{(\alpha_T - \alpha_R)\Delta T}{1 + \alpha_R \Delta T} \right] + \left[\frac{\phi_T - \phi_R}{1 + \phi_R} \right] \right). \tag{2.2}$$

In this case, ϕ_T and ϕ_R are the chemical shrinkage components in tangential and radial direction. These two values can be assumed with mixed rules if the shrinkage of the fibre (usually zero) and the matrix are known. This approach is limited to sections which have uniform orthotropic material properties throughout the thickness. For the composite angle with a different stacking Darrow [7] developed a method with equivalent thermal expansion coefficient α_{eq} and defined the sum of thermal α_{th} and chemical shrinkage α_{che} according to the following equation

$$\alpha_{eq} = \alpha_{th} + \alpha_{che} = \alpha_{th} + \frac{\Delta V}{3\Delta T}. \tag{2.3}$$

Assuming a stress-free temperature in a cure cycle the residual deformations are calculated by cooling down from the cure temperature to room temperature. This can be done in a linear static analysis. Elastic models are fast and were used by Harper and Weitsmann [8], Loos and Springer [9], Stango and Wang [10], Nelson and Cairns [11], Spröwitz et al. [12], Fernelund et al. [13]. Most of the time these elastic models are not enough sophisticated to capture the complexity of the problem and have less accuracy in order to reliable results.

Incrementally elastic models
The following approaches are models in which the modulus of elasticity $E_{(T,p)}$ changes as a function of temperature T and degree of cure p, expressed as

$$\sigma_{(t)} = \int_0^t E_{(T,p)} \frac{d\varepsilon}{dt} \, dt. \tag{2.4}$$

The first step in nonlinear transient calculation was performed by Bogetti and Gillespie [14]. They used an approach by coupling the development of a resin young modulus $E_{m(p)}$ to the degree of cure p. In this case E_m^∞ is the cured resin modulus

$$E_{m(p)} = E_m^\infty \, p. \tag{2.5}$$

As enhancement of the incremental linear elastic approach a nonlinear dependency on the degree of cure was published by White and Hahn [15]. In this case p^* describe the gel point. Below the gel point the transverse modulus E_{22} is defined constant E_m^0. Above the

9

gel point, the modulus is coupled to the degree of cure by second order polynomial with the fitting parameters a_n

$$E_{22} = E_m^0 \ , \qquad\qquad\qquad 0 \le p \le p^* \ ,$$

$$E_{22} = a_0 + a_1 p + a_2 p^2 \ , \qquad p^* \le p \quad . \qquad\qquad\qquad (2.6)$$

The modulus in fibre direction and Poission's ratio are assumed to be linearly dependent on the degree of cure, modelled by experimental testing (E_{11f} fully cured longitudinal modulus, E_{11i} uncured modulus):

$$E_{11_\alpha} = E_{11i} + \left(E_{11f} - E_{11i} \right) p \ ,$$

$$\nu_{12_\alpha} = \nu_{12i} + \left(\nu_{12f} - \nu_{12i} \right) p \quad . \qquad\qquad\qquad (2.7)$$

Later on, Johnston et al. [4] proposed a so-called CHILE model, a Cure Hardening Instantaneously Linear Elastic model, in which the modulus dependents on the temperature and the degree of cure. Johnston et al. are using the DiBenedetto equation to couple a modified glass transition temperature T^* to the degree of cure also with some fitting parameters T_n^*:

$$E_{m(T,p)}^* = E_m^0, \qquad\qquad\qquad\qquad T^* < T_{C1}^* \ ,$$

$$E_{m(T,p)}^* = E_m^0 + \frac{T^* - T_{C1}^*}{(T_{C2}^* - T_{C1}^*)} \left(E_m^\infty - E_m^0 \right), \qquad T_{C1}^* < T^* < T_{C2}^*,$$

$$E_{m(T,p)}^* = E_m^\infty \ , \qquad\qquad\qquad\qquad T^* > T_{C2}^* \ ,$$

$$T^* = \left(T_g^0 + a_{tg} \, p \right) - T, \quad T_{C1}^* = T_{C1a}^* + T_{C1b}^* \, T \ . \qquad (2.8)$$

Using a comparison between glass transition temperature and process temperature, Johnston et al. are able to identify the vitrification point in a proper way. They are the first scientists who were modelling the three phases of material conversion using two constant values for the liquid (E_m^0) and solid phase (E_m^∞) and a linear function for the transition from liquid to solid state. The resin modulus is modelled as a linear function in the rubbery state. The influence of the temperature on the modulus $E_{m(T,p)}^\infty$ was defined as simple gradient a_r

$$E_{m(T,p)}^\infty = E_{m(T,p)}^* \left(1 + a_r (T - T_0) \right). \qquad\qquad\qquad (2.9)$$

Later, Fernlund et al. [15] used the CHILE model for many different studies and it has been shown that the model provides sufficient result. To complete the review of incremental linear elastic models, a last nonlinear approach from Msallem et al. [17] is presented:

$$E^*_{m(T,p)} = E'_{posgel(T)} - \left(E'_{cure(T)} - E'_{posgel(T)}\right) \frac{\left(p^2 - p^2_{posgel}\right)}{\left(p^2_{max} - p^2_{posgel}\right)} f_{(Tg)} \quad,$$

$$f_{(Tg)} = C_3\, e^{\left(C_4 \frac{T_g - T}{C_5 | T_g - T|}\right)}. \tag{2.10}$$

In this case the resin in the rubber part varies nonlinearly by the function $f_{(Tg)}$ in equation 2.10. This is found by rheological measurements. An investigation of this issue, by the author, is presented in detail in this thesis (chapter 3.7). In the following Figure 6, a schematic overview about the cure dependency of presented approaches is given.

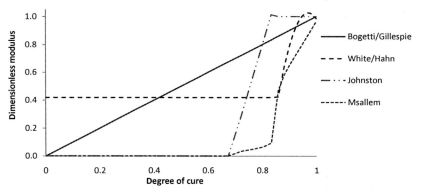

Figure 6 Cure dependent incremental elastic material models [4]

Viscoelastic models

In general, viscoelasticity means that the stress – strain relation is not linear and additional time, temperature or strain rate dependency has to be added to the elastic behaviour. For composite processing residual stress can decrease by relaxation dependent on time, temperature and curing during the process. Therefore to determine the right residual stresses the viscoelastic effect is important.

Overall, viscoelastic material behaviour can be classified in linear and nonlinear viscoelastic approaches, dependent on whether the load level will have an influence on the viscoelastic behaviour or not. Another classification can be found by the temperature dependency. It can be sorted into thermo rheological simple or complex materials. A thermo rheological simple definition means that the behaviour at high temperature for small time periods is equal to the behaviour at low temperatures for long time periods. For application to composite processing, approaches with a thermo chemical dependency are available. In these models, the viscoelastic behaviour is additionally dependent on the degree of polymerisation.

The modelling of viscoelastic behaviour of polymer materials last over four decades and the existing models are manifold. Early models are treating the viscoelastic behaviour only with temperature changes. The dependency of the viscoelasticity on the degree of cure was first modelled by White and Hahn [19]. These ideas have been used and enhanced by different researchers. Most of these approaches are using the assumption of linear viscoelasticity. In the following, firstly the general mathematical theories of linear viscoelasticity are presented and secondly, some of the most advanced and complex theories of viscoelasticity are presented.

Viscoelastic models can be written in integral or differential forms. In most publications, researchers use the integral representation. For a 1-D dependency, in case of isothermal conditions, this can be written as follows whereat $E_{(t-\tau)}$ is defined as the viscoelastic modulus, ε is the strain, t is the current time and τ the time integration variable

$$\sigma_{(t)} = \int_0^t E_{(t-\tau)} \frac{d\varepsilon}{\partial \tau} \partial \tau .$$

(2.11)

Different researchers like Johnston et al. [4] have simplified the temperature dependency on the Young's modulus using linear gradients. A more advanced approach is the application of the Time-Temperature Superposition (TTS) principle. In this case, the constitutive equation takes the following form

$$\sigma_{(t)} = \int_0^t E_{(t-\tau,T)} \frac{d\varepsilon}{\partial \tau} \partial \tau .$$

(2.12)

Different approaches by Svanberg et al. [21], Blumenstock [22] have used the approach of Schapery [23] to define the material as thermo rheological simple. The underlying assumption for this approach will be that the material behaves thermo-rheological simple and the effects of time and temperature are similar. In this case, the equation can be simplified by introducing reduced-time formulation $\xi_{(t)}$ and $\xi'_{(\tau)}$ using a time temperature shift factor a_T :

$$\sigma_{(t)} = \int_0^t E_{(\xi-\xi')} \frac{d\varepsilon}{\partial \tau} \partial \tau ,$$

(2.13)

$$\xi_{(t)} = \int_0^t \frac{1}{a_T} dt , \quad \xi'_{(\tau)} = \int_0^\tau \frac{1}{a_T} dt .$$

(2.14)

A commonly used method to represent viscoelastic materials in the differential form are spring and damping elements. A simple standard form of these representations is the so called Maxwell element, a combination of a spring and a damper element in series, and the Kelvin element, a combination of a spring and a damper element in parallel. In most publications, researchers such as Wijskamp [24] have extended the micro mechanical model of Voigt and Reuss [54] by adding a damper element (Fig. 7). Therefore the matrix was taken as a Maxwell element.

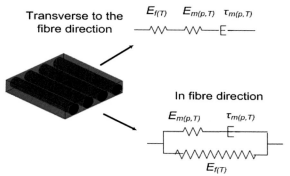

Figure 7 Schematic material model

To clarify the different viscoelastic approaches in the next section, two of the most advanced models are presented in details which add a cure dependency to the viscoelastic behaviour. Zobeiry [16] defines them as a "gold standard" of viscoelastic models for composite processing with a time, temperature and polymerisation dependency. In general, the material behaviour can be modelled by using a discrete approach such as a Prony series or a closed function using a mathematical expression. The first model of Kim and White [26] take a reduced time formulation $E_{(\xi-\xi')}$ and a number of Maxwell elements based on the characteristic relaxation time $\tau_{\omega(p)}$, idealised with a polynomial function $f'_{(p)}$ to implement the cure dependency (2.15). The second model, published by Prasataya et al. [27], based on a bulk modulus definition $K_{(t,T,p)}$. The cure dependency is defined using the glass transition temperature based on the Williams Landel Ferry (WLF) equation (2.16)

Model of Kim and White [26]

$$\sigma_{(t)} = \int_0^t E_{(\xi-\xi')} \frac{d\varepsilon}{\partial\tau} \partial\tau \,,$$

$$E_{(p,\xi)} = E^\infty_{m(p)} + \left(E^u_{m(p)} - E^\infty_{m(p)}\right) \sum_{\omega=1}^{n} w_{\omega(p)} e^{\left(-\frac{\xi_{(p,T)}}{\tau_{\omega(p)}}\right)},$$

$$log_{(\tau_{\omega(p)})} = log_{(\tau_{\omega(p^0)})} + \left[f'_{(p)} - (p - p^0) log\left(\lambda'_\omega\right)\right] \,,$$

$$f'_{(p)} = -9{,}3494 + 0.6089p + 9.1347p^2,$$

$$\lambda'_\omega = \frac{\tau_{p(p^0)}}{\tau_{\omega(p^0)}},$$

$$\xi = \int_0^t \frac{1}{a_T} dt,$$

$$log_{(a_{T,p})} = c_{1(p)}T + c_{2(p)},$$

$$c_{1(p)} = -a_1 e^{\frac{1}{p-1}} - a_2, \qquad c_{2(p)} = -T^0 c_{1(p)}, \tag{2.15}$$

Model of Prasataya et al. [27]

$$K_{(t,T,p)} = K_r + (K_g - K_r) \sum_{i=1}^{n} g_i e^{\left(-\frac{t}{a_{T,p}\tau_i}\right)},$$

$$log_{(a_{T,p})} = \left(\frac{C}{T - T_\infty} - \frac{C}{T_{ref} - T_\infty}\right) - \left(\frac{C}{T_{(g(p))} - T_\infty} - \frac{C}{T_{g(p_{ref})} - T_\infty}\right). \tag{2.16}$$

As shown in these examples, the extension of the time-temperature viscoelastic behaviour to a time-temperature-cure dependency is done by adding more complexity to the shift factor a_T to $a_{T,p}$. Both models are describing the viscoelastic behaviour of an isotropic polymer and are not applicable for orthotropic composite materials but the idea of complex shift factors which based on time, cure, temperature and additional parameters like material direction or fibre volume content is promising.

For orthotropic or anisotropic reinforced composite materials the viscoelastic behaviour is strongly influenced by the reinforcements, as shown in Figure 7. Accordingly an approach is needed which enables to introduce different viscoelastic behaviour based on the material directions. Zocher et al. [28] has published an incremental definition for an orthotropic material and showed for different generic 2D problems reliable results. This theoretical work of Zocher et al. [28] is used by Svanberg et al. [25] and simplified / adapted to describe the viscoelastic behaviour of an orthotropic composite material. Therefore, to define the stress tensor at the actual time step $\sigma_{ij\,(t+\Delta t)}$, an incremental form has to be taken by adding the present stress state $\sigma_{ij\,(t)}$ on the actual stress increment $\Delta\sigma_{ij}$ which is defined by equation 2.18.

$$\sigma_{ij\,(t+\Delta t)} = \sigma_{ij\,(t)} + \Delta\sigma_{ij}, \tag{2.17}$$

$$\Delta\sigma_{ij} = \Delta\sigma_{ij}^{R} + \Delta C_{ijkl} \cdot \Delta\varepsilon_{kl}. \tag{2.18}$$

By adding series of parallel Maxwell elements, also known as Generalized Maxwell Model or Maxwell–Wiechert Model, on the incremental stiffness tensor ΔC_{ijkl} and the stress increment $\Delta\sigma_{ij}$ the viscoelastic behaviour can be implemented

$$\Delta C_{ijkl} = C_{ijkl}^{\infty} + \frac{1}{\Delta\xi} \sum_{p=1}^{n} \rho_{ijkl}^{p}\, C_{ijkl}^{p} \left(1 - e^{\frac{-\Delta\xi}{\rho_{ijkl}^{p}}}\right), \tag{2.19}$$

$$\Delta\sigma_{ij}^{R} = -\sum_{k=1}^{3}\sum_{l=1}^{3}\sum_{p=1}^{n} S_{ijkl\,(t)}^{p} \cdot \left(1 - e^{\frac{-\Delta\xi}{\rho_{ijkl}^{p}}}\right). \tag{2.20}$$

Svanberg used $S_{(t+\Delta t)}$ as recursive element for relaxation of $\Delta\sigma_{ij}^{R}$

$$S_{(t+\Delta t)} = e^{\frac{-\xi}{\rho_{ijkl}^{p}}} \cdot S_{ijkl\,(t)}^{p} + C_{ijkl}^{p}\, \frac{\rho_{ijkl}^{p} \cdot \Delta\varepsilon_{total}}{\Delta\xi} \left(1 - e^{\frac{-\Delta\xi}{\rho_{ijkl}^{p}}}\right). \tag{2.21}$$

The relaxation behaviour is based on the parameters ρ_{ijkl}^{p} and $\Delta\xi$. As shown in the previous section, different researchers like Kim and White [26], Prasataya et al. [27], Zobeiry [16] have added a dependency of time, temperature, polymerisation by modifying the shift factor a_T. In the published work of Svanberg et al. [25] the following simplified shift factor a_T definition can be found and additionally a simplification of the Generalized Maxwell element was done using McLaurin series

$$\Delta\xi = \frac{\Delta t}{a_T}. \tag{2.22}$$

Svanberg et al. [21] concern the following citation about the shift factor "To obtain a material model that is simple but still capture most of the important mechanisms for a curing problem we approximate the time-cure-temperature shift factor in the following manner". They simplified the relaxation spectrum to one constant ω and take the inverse based on the definition of the vitrification point using the glass transition temperature T_g

$$a_T = \lim_{\omega \to 0} \begin{array}{ll} \omega, & T > T_g, \\ \frac{1}{\omega}, & T < T_g. \end{array} \tag{2.23}$$

Svanberg et al. [25] have not introduced a viscoelastic behaviour which takes into account the orthotropic material behaviour and they have not presented any experimental characterisation to validate their approach. Accordingly the reliability of the presented approach has to be proven against experimental characterisation using relaxation tests. Additionally, more complex shift factors could to be used to take into account the viscoelastic behaviour based on time, cure, temperature, material direction and / or fibre volume content.

Recapitulatory, there are three classes of material model: elastic models, incremental elastic models and viscoelastic models. They differ in their capability to capture the complexity of the problem. This means adding complexity to the model increases accuracy but also computational and characterisation effort. Consequently, sophisticated idealisations and simplifications are needed to take into account all necessary physical effects. Therefore a viscoelastic model which is computational efficient and easy to characterise is needed.

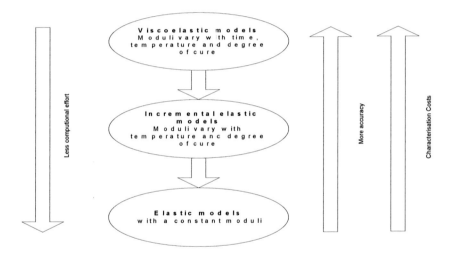

Figure 8 Different levels of material models [16]

3 Material characterisation – Single Components

This chapter addresses the characterisation of the thermo-mechanical behaviour of resin and fibres as single components, subject to the curing reaction and temperature. In chapter 2.2 the demands on the simulation approach have been defined to represent the thermo-chemical and thermo-mechanical behaviour. Consequently, characterisation and modelling of relevant thermo-chemical parameters like curing reaction, heat of reaction, heat capacity, thermal conductivity and density is needed. Furthermore, characterisation and modelling is required for mechanical parameters like modulus, thermal contraction, chemical induced shrinkage, relaxation etc. Key aspects are the development of the resin properties like Young's modulus, thermal expansion, chemical shrinkage, relaxation behaviour dependent on the degree of cure and temperature. The chapter is structured in the way that the single physical quantities are separately analysed. Characterisation of the single key aspects ends up with constitutive equations and parameters for the numerical analysis. A unidirectional composite (G1157) consisting of the C-Fibre TENAX HTA 5131 and the premixed epoxy resin HEXCEL RTM 6 is used for all applications and validation test in this thesis.

The carbon fibre properties as shown in table 3.1 are based on a product of TENAX HTA 5131 fibers which are, for example, part of G1157 D 1300 Injectex fabric. The type of fibre is part of the high tenacity class [29].

Table 3.1 Fibre properties [54]

Young's modulus in fibre direction	E_{f1}	$= 210000\,N/mm^2$
Young's modulus transverse	E_{f2}	$= 28240\,N/mm^2$
Shear modulus	G_f	$= 50600\,N/mm^2$
Poisson's ratio	υ_{f12}	$= 0.225$
CTE in fibre direction	α_{f1}	$= -0.045 * 10^{-6}\,1/K$
CTE transverse to the fibre direction	α_{f2}	$= 12.5 * 10^{-6}\,1/K$
Strength in fibre direction	R_f	$= 3430\,N/mm^2$
Fibre diameter	d_f	$= 7\mu m$

The presented study concentrates on a polyfunctional epoxy resin, RTM 6, supplied by HEXCEL Composites (UK). The resin, intended for the manufacturing of aircraft composite structures, using the resin transfer moulding (RTM) process, is currently receiving considerable attention from civil aircraft manufacturers. The expected service temperatures of final products range from -60°C to 180°C. The uncured resin has a density of 1.117 g/cm³ and the fully cured resin has a density of 1.141 g/cm³. The recommended cure regime for this resin is 160°C for 75 min, followed by a post-cure at 180°C for 2 hours, resulting in an expected glass transition temperature of about 183°C in the cured material. Alternatively, five different cure cycles are possible and approved by the supplier. In this study all samples are manufactured using cycle number five [30]. In this case the recommended curing temperature of 180°C for 120 min is resulting in a glass transition temperature of

196°C and a final degree of cure by 96%. The following resin properties are given for the cured material [30].

Table 3.2 Matrix properties (cured) [30]

Young's modulus	E_m	$= 2890 \, N/mm^2$
Shear modulus	G_m	$= 1070 \, N/mm^2$
Poisson's ratio	v_m	$= 0.35$
CTE	α_m	$= 65 * 10^{-6} \, 1/K$
Strength	R_m	$= 75 \, N/mm^2$

3.1 Cure kinetics

The description of the curing process of thermoset resin, which transforms from the liquid into the solid state at an elevated temperature, can be described using generalised empirical rate equations or mechanistic models. Mechanistic models start to describe the curing process from an atomistic scale by declination of the growing macromolecules. The empirical rate approach derives a phenomenological mathematical formulation of the curing process by experimental studies using Differential Scanning Calorimetry (DSC). Using this method the exothermic heat flow is measured and interpreted upon the assumption, that this heat flow is proportional to the degree of cure. The reaction kinetics is used to describe a reaction of n-th order with p representing the conversion factor (varying from 1 to 0) also known as the degree of cure. The temperature dependency is described by an Arrhenius rate constant k_i:

$$\frac{dp}{dt} = k_1 \cdot p^n, \qquad k_i = A_i \cdot e^{\left(\frac{S_i}{RT}\right)} \quad i = 1 \dots 3 \qquad (3.1)$$

The Arrhenius term is based on material and process parameters with the thermal activation energy of reaction S_i, universal gas constant R, and temperature T. The variety of kinetic reaction models available in the literature for different resin systems is manifold. In general, the model must be able to handle different cure situations, especially taking into account the variation of isothermal and dynamic temperature conditions.

Kamal and Sourour developed the following model which is mostly used for the given resin system (RTM6) in the literature [31]

$$\frac{dp}{dt} = (k_1 + k_2 \cdot p^m) \cdot (1 - p)^n \,. \qquad (3.2)$$

As shown in Fig. 9 the development of the degree of cure using the approach of Kamal / Sourour agrees with experimental measurements in a temperature range only from 160°C to 180°C. In case of lower curing temperatures this approach does not provide reliable results and will explained in the following.

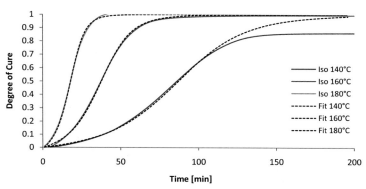

Figure 9 DSC results and analytical fit for different isothermal curing conditions of pure RTM6 resin

Ryan and Dutta [32] extended the model and varied the variables m and n depending on the temperature. This might help to increase accuracy of the approach but the reaction kinetic describes only the chemically controlled curing. The reaction is treated, as being influenced by the temperature and the chemical formulation. This assumption is not valid if a three dimensional solid network is formed. After reaching the vitrification point the curing will be controlled by diffusion. If the curing progresses, the solid network will limit the probability of finding possible reaction partners. There are currently three methods for adding a rate-limiting diffusion term to the cure rate equation: first, an expression for the diffusion rate from Rabinowith [33] based on the free volume theory, second, a method which is modifying the definition of the final degree of cure and third, use of a diffusion factor based on the glass transition temperature.

Approach based on the final degree of cure p_{End}:

$$f_{D(p)} = \left[\frac{2}{1 - e^{[(\frac{p - p_{End}}{b})]}} - 1 \right], \tag{3.3}$$

$$\frac{dp}{dt} = f_{(p)} \cdot f_{D(p)}. \tag{3.4}$$

Approach based on the glass transition temperature [33] :

$$k_D = e^{\left(\frac{-b}{0.00048(T - T_G) + 0.025} \right)}, \tag{3.5}$$

$$\frac{dp}{dt} = \left(\frac{k_1 k_D}{k_1 + k_D} + \frac{k_2 k_D}{k_2 + k_D} \cdot p^m \right) \cdot (1 - p)^n. \tag{3.6}$$

Especially for low curing temperatures, a diffusion term extension must be used. In Fig. 10 the Kamal / Sourour model is extended with the diffusion factor approach. The different theories have been proved and it was found that the approach using the diffusion factor is useful and only one parameter *b* has to be needed.

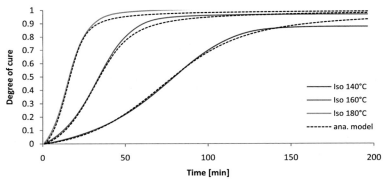

Figure 10 DSC results and analytical fit for different isothermal curing conditions including diffusion factor (RTM6 + binder)

To find the cure kinetic parameters a detailed study using dynamic curing test (heating rate 20°C/min, 15°C/min, 10°C/min, 7.5°C/min, 5°C/min, 2°C/min) and isothermal curing test (120°C, 140°C, 160°C, 180°C) have been performed on pure RTM6 resin and RTM6 resin with binder. The binder EPR 05311 (HEXCEL) is added on the fabric to enhance performance of the textiles with a proportion of around 5% on the resin mass. The total mass of the binder depends on the preform technology and can vary local [34]. The binder is based on a thermoset system with less reactive groups. Therefore, there is an influence on the reaction and a different set of cure kinetics parameters have to be used. The results of the detailed study can be found in Fact / Vitech WP 4.1, [35, S3]. Table 3.3 summarises the results of this detailed study:

Table 3.3 Parameters	RTM6		RTM6 + Binder (5%)	
Reaction enthalpy	H_{tot}	$= 419 \, J/g$; St. Dev. 3%	H_{tot}	$= 400 \, J/g$ St. Dev. 3.5%
Arrhenius term	A_1	$= 4.5 \cdot 10^6 \, /sec$	A_1	$= 8.0 \cdot 10^6 \, /sec$
Arrhenius term	A_2	$= 1.3 \cdot 10^6 \, /sec$	A_2	$= 0.8 \cdot 10^6 \, /sec$
Activation energy	S_1	$= 74690 \, J/mol$	S_1	$= 76690 \, J/mol$
Activation energy	S_2	$= 58370 \, J/mol$	S_2	$= 58180 \, J/mol$
Reaction order	n	$= 1.15$	n	$= 1.00$
Reaction order	m	$= 1.20$	m	$= 0.95$
Diffusion factor	b	$= 0.05$	b	$= 0.12$

The conclusion of the cure kinetic study is that the approach of Kamal / Sourour with an extension for the diffusion behaviour is a good approximation in case of isothermal curing conditions. The influence of the binder is less but should be taken into account, especially, if the binder content is higher than 5%.

3.2 Glass transition temperature

During the cure of epoxy resin, three different morphologic states are passed through. The resin converts from a liquid to a rubbery, and then to solid state. These changes are, of course, not defined as points, but from the engineering point of view they will be defined as points and can be determined in connection to the glass transition temperature. The change of the glass transition temperature is dependent on the degree of cure. This can be taken into account by the following DiBenedetto and Venditti-Gillham equation available in the literature [36, 53]

DiBenedetto equation

$$\frac{T_g - T_{g0}}{T_{g1} - T_{g0}} = \frac{\lambda_{Tg} \cdot p}{1 - (1 - \lambda_{Tg}) \cdot p} \quad , \tag{3.7}$$

Venditti-Gillham equation

$$\frac{lnT_g - lnT_{g0}}{lnT_{g1} - lnT_{g0}} = \frac{\lambda_{Tg} \cdot p}{1 - (1 - \lambda_{Tg})p} \quad . \tag{3.8}$$

There are four important points during the cure. Below T_{g0} there is no reaction able to start. The degree of cure is 0. This temperature is usually below room temperature. During low temperature the resin is not able to start curing because the internal energy is too low to start the autocatalytic process. If the gel point T_{gel} is reached, a first three-dimensional network exists. This temperature defines the point where the liquid state transforms to the rubbery state. The next important point is defined where the glass transition temperature is higher than the cure temperature. This transition point is known as vitrification and the associated loss of mobility means that the reaction starts to be more or less diffusion controlled. The rubbery resin will transform to a glassy solid state and the mobility of the reacting groups is decreasing. The last important temperature is defined by the maximal glass transition temperature T_{g1} where the resin is fully cured (degree of cure 1). Above this temperature there will be no additional curing. In most cases for RTM6 resin the DiBenedetto equation is used [37, S10].

From a study which was performed by Dykemann [36], it is shown that the glass transition – degree of cure behaviour is dependent on isothermal and dynamic temperature variations. It is mentioned that in case of increasing temperature, the glass transition temperature T_g is also increasing. To capture this effect the DiBenedetto equation was combined with an S-shape curve

$$T_g = \frac{\lambda_{Tg} \cdot p (T_{g1} - T_{g0})}{1 - (1 - \lambda_{Tg}) \cdot p} + T_{g0} + \frac{D}{1 + e^{-F(\alpha - \alpha_{crit})}} \quad . \tag{3.9}$$

In this extension, D is a multiplier with the value of 35°K, F is the breadth of the transition and a value of 25°K is mentioned to match the transition for isothermal and dynamic cure. α_{crit} is the mid-point of the S-shape transition and was found to change with temperature and heating rate. To determine the right parameters of the glass transition temperature modulated DSC experiments have been performed. To investigate the development of the glass transition temperature depending on degree of cure, the DSC was heated up to 120°C, 140°C, 160°C, 180°C, 210°C, 250°C and 320°C, between the different temperatures every time it was cooled down to -50°C. The glass transition was measured by the change of the specific heat capacity. The results are shown in the following Figure 11.

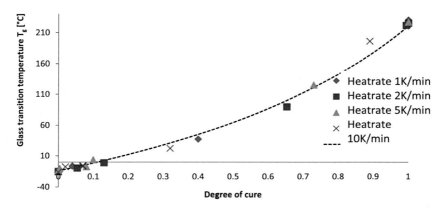

Figure 11 DSC results and analytical fit for different dynamic curing conditions

Analysing the experiments, a good correlation of the DiBenedetto with the extension of Dykemann [36] was found (Fig. 11). The following parameters are the results of the experimental study:

Table 3.4 Parameters

Fit parameter	λ_{Tg} = 0.50
Lowest Temperature	T_{g0} = $-15\ [°C]$
Max. Temperature	T_{g1} = 170 [°C]
Fit parameter	D = 50 [°C]
Fit parameter	F = 52 [°C]

3.3 Density

The density which accrues in the heat transfer equation is influenced by chemical and thermal shrinkage during the process. Figure 12 demonstrates schematically the change of the scalar quantity [45]. During the heating up the resin expands (A-B), during the curing the chemical shrinkage is active (B-C) and ends up with additional constriction during cooling down (C-D-E). The total shrinkage is a sum of thermal and chemical shrinkage and can be determined easily by measuring the density of an uncured and a cured material.

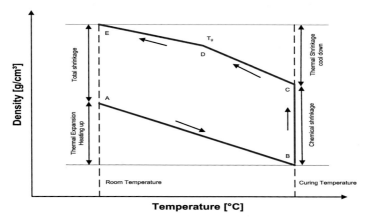

Figure 12 Density changes during process [45]

For the studied material, RTM6, the following information are available from product data sheet [30]

uncured resin $\quad \rho_{m,25°C} = 1.11\frac{g}{cm^2}$,

cured resin $\quad \rho_{m,25°C} = 1.14\frac{g}{cm^2}$. $\qquad$ (3.10)

In this work no experimental density measurements are performed, but it has to be mentioned that different researchers like Mijovic and Wang [41], Lam and Piggott [42], Armstrong et al. [43] have performed studies on the interrelationship between density, temperature or/and degree of cure. Johnston et al. [4] published an approach to incorporate the change of the density dependent on the degree of cure and temperature

$$\rho_{m(p,T)} = \rho_{m0} + a_\rho(T - T_0) + b_\rho(p - p_0) . \qquad (3.11)$$

3.4 Specific heat capacity

The variation of the heat capacity with degree of cure and reaction temperature is one of the sensitive parameters to describe the thermal behaviour of thermoset resin. The specific heat will increase with the temperature, but it decreases changing from liquid to a solid state. This dependency is not linear. The change of the specific heat capacity between liquid to solid will be approximately 15 to 20% and leads to a direct error in the idealisation of the thermal material behaviour. Consequently, the dependency of the specific heat capacity on temperature and degree of cure has to be considered.

There are different approaches to idealise cure dependencies; one comes from Balvers [38], who proposes to define $c_{p,m}$ by a hyperbolic function depending on the temperature and the glass transition temperature:

$$c_{p,m} = a_1 T + a_2 + a_5(T - T_g) + (a_3 - a_5 T_g) tanh \ \left(m^- a_4(T - T_g)\right), T \leq T_g$$

$$c_{p,m} = a_1 T + a_2 + a_5(T - T_g) + (a_3 - a_5 T_g) tanh \ \left(m^+ a_4(T - T_g)\right), T > T_g$$

$$m^+ = \frac{a_5 + F}{F}, \qquad m^- = \frac{-a_5 + F}{F},$$

$$F = a_4(a_3 - a_0 T_g). \tag{3.12}$$

For this approach, five parameters have to be evaluated. A second approach was published by Chern et al. [46], using a simpler definition. In this issue $c_{p,m}$ depends only on the temperature and directly on the degree of cure:

$$c_{p,m} = a_1 + a_2 T - a_3 p \ , \qquad\qquad T \leq (208.8 + 293.26 * p)$$

$$c_{p,m} = a_4 + a_5 T - a_6 T^2 - a_7 T^3 \quad . \qquad T > (208.8 + 293.26 * p) \tag{3.13}$$

In this case, seven material parameters a_n have to be evaluated. The third approach which can be found by Johnston et al. [4] idealised the behaviour similar to the work of Balvers. Johnston et al. [4] define that in the solid phase the heat capacity is only dependent on temperature. In the rubbery phase $T \leq T_g$ the behaviour is also depending on the degree of cure:

$$c_{p,m} = a_1 + a_2 T + a_3 p \ , \qquad\qquad T \leq T_g$$

$$c_{p,m} = a_4 - a_5 T. \qquad\qquad\qquad T > T_g \tag{3.14}$$

Similar to the approach of Balvers, five parameters a_n have to be evaluated. In a separated study [35], a comparison between the three different approaches was done for an isothermal curing condition at 180°C. It was visible that for isothermal curing condition all three approaches are showing a similar behaviour but for non-isothermal conditions they differ largely. For determining the specific heat capacity of RTM6 resin as a function of degree of cure and temperature, isothermal and dynamic Modulated Differential Scanning Calorimetry (MDSC) runs have been carried out. MDSC means that the temperature is modulating with temperature amplitude of 0.5°C and a period of 60s to obtain reversible and non-reversible heat flow. The reversible heat flow is used for determine the specific heat capacity.

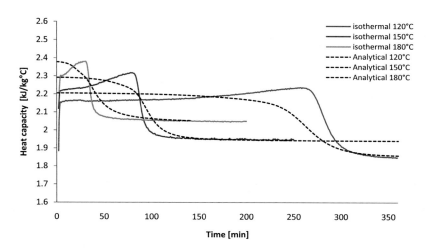

Figure 13 Changes of the heat capacity during isothermal curing

In Fig. 13 the measured values are compared to the approach of Balvers. This is a result of a detailed comparison between the three formulations. A best correlation was found between experimental values and modelling using the eq. 3.12 developed by Balvers. The following tables summarise the determined parameters.

Table 3.5 Parameters

Fit parameter	$a_1 = 0.00264$
Fit parameter	$a_2 = 1.1$
Fit parameter	$a_3 = 0.172$
Fit parameter	$a_4 = 0.0423$
Fit parameter	$a_5 = 0.000242$

In general, such a complex formulation of the heat capacity is questionable and in the literature different statements can be found. Balvers [38] has summarised following statement: "From the results it can be concluded that taking a constant value, which corresponds to the initial reaction temperature and degree of conversion, is in good agreement with the simulation, in which the matrix's specific heat capacity is a function of reaction temperature and degree of conversion". Nevertheless the change of the heat capacity dependent on temperature and degree of cure is not small and consequently the approach of Balvers with the derived parameters are used in this work.

3.5 Thermal conductivity

The thermal conductivity change, influenced by the temperature and polymerisation, similarly to the heat capacity, but it is noteworthy that the thermal conductivity is a vector quantity. In the case of pure resin, an isotropic material behaviour can be defined and all thermal conductivity vector values can be equalised. There are following approaches to model the thermal conductivity, proposed by Skordos et al. [47]. He defines the thermal conductivity in respect to the temperature and the degree of cure in the following way

$$k_{c,m} = b_1 T p^2 - b_2 T p - b_3 T - b_4 p^2 + b_5 p + b_6 \ . \tag{3.15}$$

Alternatively, Chern et al. [48] presents a formulation without a dependency of the degree of cure on the thermal conductivity. In this paper, the behaviour of a Hercules 3501-6 epoxy resin was investigated. The result of this study was that the thermal conductivity is only weakly deponent on the degree of cure. It is idealised with an average temperature dependent thermal conductivity by a fourth order polynomial

$$k_{c,m} = b_1 + b_2 T - b_3 T^2 + b_4 T^3 - b_5 T^4 \ . \tag{3.16}$$

Also by Johnston et al. [4], an approach can be found to model the thermal conductivity in dependency on the degree of cure and temperature

$$k_{c,m} = b_1 + b_2 T + b_3 p \ . \tag{3.17}$$

In comparison to all three approaches, Johnston and Chern provide more reliable trends. The approach of Skordes was not able to show good results for the RTM6 resin [35]. The approach of Johnston et al. was, compared to the approach of Chern, simpler with less parameter. In this thesis, measurements of the conductivity during curing have been performed. These measurements were not successful, in view of the exothermal heat which avoided reaching isothermal conditions [S11]. Additionally an experimental measurement of the thermal conductivity has been performed on cured resin samples using Laser Flash Analysis (LFA) method in the temperature interval of RT to 200°C. The results of the process temperature range are presented in Figure 14.

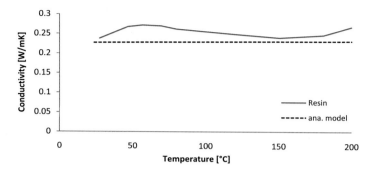

Figure 14 Temperature dependency of the conductivity of cured resin

Accordingly, the approach of Johnston [4] was used with the following values.

Table 3.6 Parameters

Fit parameter	$b_1 = 0.15 \, [\text{W/mK}]$
Fit parameter	$b_2 = 0.0000245 \, [1/°\text{K}]$
Fit parameter	$b_3 = 0.07$

Accordingly, it could be conclude that temperature dependency is less for the given resin system and dependency on the degree of cure was not possible to measure. Nevertheless the chosen approach of Johnston allowed take into account both dependencies. In the given case the value a_3 was taken by the literature reference of Johnston [4].

3.6 Chemical Shrinkage

The chemically induced volumetric shrinkage is directly related to the degree of cure and is one of the main reasons for distortions and residual stresses. Consequently, the total amount of chemical induced shrinkage and the cure related behaviour is important in order to understand the development of residual stresses. In general, two approaches are available to determine shrinkage: volume dilatometry, here the change of a dimension is measured, and non-volume dilatometric methods, here the shrinkage is derived by equivalent variables, for example warpage of a two-layer plate ([0,90] layup), force, strain of fibre bragg grating sensors, etc. It can be concluded that these two different approaches lead to different results related to the fact that non-volume dilatometry is problematic, because the contact between resin and a solid surface (mould) can induce stresses. Therefore, different statements can be found in the literature about the dependency of chemical induced shrinkage on the degree of cure. Holst [45] defines a linear dependency (volumetric dilatometric method), while Liu [49] constitute bilinear behaviour (non-volumetric dilatometric method).

Using a Dynamic Thermo Mechanical Analysis (DTMA) setup, a simple test using a volume dilatometer method has been performed. Encapsulated DSC pans have been filled with resin and placed inside the DTMA compression test setup. Applying a small preload force, the deformation of the cap during isothermal curing conditions at 150°C has been observed. The results gained using the analytical approach of Kamal / Sourour [31] are plotted versus the degree of cure in Figure 15.

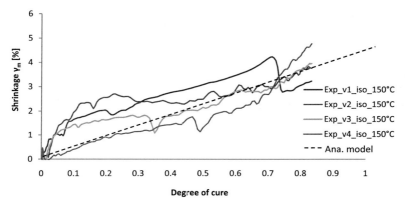

Figure 15 Resin shrinkage against degree of cure, measured using TMA setup

The result of this experiment will be that sources of error like friction between the cap and the pan, overflow of the resin, and the preload force have a considerable influence. This can be observed in the large variation of the measurement values. Nevertheless, a trend is visible that the shrinkage of the resin can be idealised as a linear function of the degree of cure. The total amount of chemical volumetric shrinkage V_{sh} was in average $\gamma_m = 4.8\%$ for a total cured state.

To compare these results with a measuring method of type non-volume dilatometry, a second measurement was performed using a rheometer. Normally, a rheometer test is used to measure the response of applied moments of a fluid between two parallel plates. Additionally, it is possible to measure a gap variation with a controlled normal force. The method is only suitable to measure shrinkage after passing the gel point. On an AR2000 rheometer from TA Instruments a plate to plate test setup with the diameter of 25mm and a force threshold of 0.1N was used during isothermal curing conditions at 150°C. The initial gap was set to 500µm. Following diagram (Fig. 16) is showing the gap variation in comparison over the curing time. On the primary axis (on the left side) the development of the real and imaginary shear modulus is plotted. Using these curves the gel point can be identified by the intersection of the G' and G'' curves. On the secondary axis (on the right side) the gap variation is displayed. It is visible that the cure-induced shrinkage starts after passing the gel point. The gap variation changes from $h_0 = 500\mu m$ to $h = 475.5\mu m$ on average.

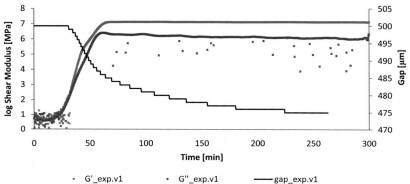

Figure 16 Rheometer measurement of the chemical shrinkage during isothermal curing at 150°C

Khoun et al. [50] have performed a similar test and derived the following formula to compute the volumetric shrinkage γ_m out of the gap displacement

$$\gamma_m = \left(1 + \frac{1}{3}\left(\frac{h - h_0}{h_0}\right)\right)^{-3} - 1. \tag{3.18}$$

Accordingly, this means a total amount of shrinkage $\gamma_m = 3.84\%$. The volumetric strain can be displayed over the degree of cure using the cure kinetic approach of Kamal / Sourour (Fig. 17). Using this non-volume dilatometric method a chemical shrinkage occurs after reaching the gel point. After this a similar linear dependency on the degree of cure can be assumed.

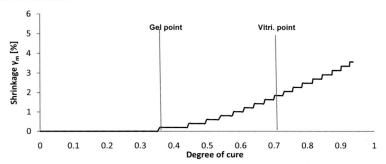

Figure 17 Chemical shrinkage over the degree of cure for isothermal curing at 150°C

In comparison to the TMA measurement method (volume dilatometry), the rheometer measurement (non-volume dilatometric method) shows more reliable results. The variation between the two measurements and the amount of uncertainties is smaller. Consequently, it is recommended to use the rheometer method for the measurement of cure shrinkage. It can be concluded that chemical shrinkage occurs after the gel point and can be assumed linear dependent on the curing reaction.

Using this result, shrinkage can be introduced into the analysis using an incremental strain formulation for isotropic materials. The total volumetric shrinkage γ_m has to be multiplied with the increment of the degree of cure to get the chemical induced shrinkage strain increment $\Delta \varepsilon_{sh(p)}$

$$\Delta \varepsilon_{sh(p)} = \frac{\gamma_m}{3} \Delta p \ . \tag{3.19}$$

3.7 Mechanical behaviour during the curing process

The constitutive model of the resin has to represent the mechanical behaviour at the different stages of the curing process. It is useful to divide the development of the resin properties into three separate phases: first, a phase of purely viscous behaviour where the resin does not support loads and does not develop residual stresses, second, a phase of viscoelastic behaviour after gelation (gel point) where stresses will be generated but decay to some degree, relative to the process time, and third, a phase of elastic behaviour after vitrification, where the material behaves nearly linearly. A material model must take into account all three stages. First formulations in the last decades started to couple the resin modulus directly to the degree of cure (Bogetti and Gillespie, [14]). Further approaches can be found in the literature on the aspect of how to model the resin modulus development by either linear, incremental linear (Johnston [4]) or non-linear approaches (White and Hahn, [19]; Msallem et al., [17]) which is schematically shown in Figure 6.

Therefore, an important objective question will be the classification of the resin development versus degree of cure. To answer this question, shear rheological measurements have been performed using a plate to plate method with a frequency of 1 Hz and an amplitude of 0.1% strain. Four test cases are investigated for isothermal conditions at temperatures 120°C, 150°C, 180°C and 200°C. The fig. 18 represents the storage and loss shear modulus displayed on a logarithmic scale over curing time. For determination of the gel point, the intersections of the storage modulus (full line) and the loss modulus (dashed line) have been used. The point of vitrification is evaluated by fulfilment of a non-change criterion of the storage modulus over time (Fig. 18).

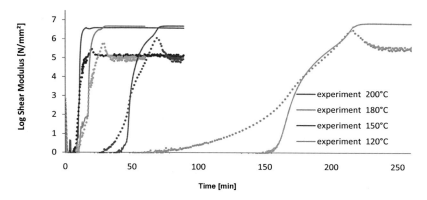

Figure 18 Development of the storage (full line) / loss (dashed line) shear modulus during different isothermal curing conditions

As a first result, the storage modulus curves versus time can be classified in the way that the development of the resin modulus for slow curing processes at low temperatures represents a nonlinear behaviour. In the case of a fast reaction at high temperatures, the development seems to be linear. If the experimental curves are plotted against the degree of cure, this has to be rectified. In all four cases, the resin modulus development must be represented, at least by an incremental approach [4] which takes into account the gel point and the point of vitrification. Using the shear rheometer experiments, the development of the resin during cure was observed at an isothermal curing temperature of 150°C. The development of the resin modulus can be idealised by the following function:

$$G = \left[\frac{1}{\left(\frac{1-\eta}{1+c(T^{**}-1)^\zeta} + \eta \right)} \right] \eta \cdot G_\infty , \tag{3.20}$$

$$T^{**} = \frac{1}{\left(1 - \dfrac{T_g - T_{gel}}{T - T_{gel}} \right)} . \tag{3.21}$$

G_∞ describes the shear modulus of the completely cured material, η defines the modulus in the liquid state and c and ζ are fitting parameters for experimental data. The advantage of this formulation is a more accurate, but also smooth function, as the modulus can be modelled in a good agreement to experimental data (Fig. 19).

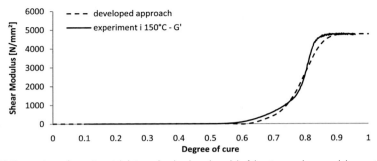

Figure 19 Comparison of experimental data and a developed model of the storage shear modulus against the degree of cure

It is easy to adapt the model to experimental results using the lower bound η, shift c and rotation ζ factors. Comparing the proposed formulation to the earlier discussed formulations (Figure 6) with respect to numerical integration schemes, the proposed formulation is also more convergence stable as the solver does not have to pass points of discontinuity as e.g. found in the formulation of Johnston et al.[4] at the gel and vitrification point (eq. 2.8).

Temperature dependency of the modulus and thermal expansion

Thermoset polymers behaviour is strongly influenced by the temperature. Accordingly, the objective question is how the Young's modulus and the coefficient of thermal expansion (CTE) of the cured resin behave in respect to temperature changes. To answer this question, a Dynamic Mechanical Thermal Analysis (DTMA) experiment has been performed to measure the Young's modulus - temperature dependency using a three point bending test setup. A square plate with the dimensions 35mm*10mm*2mm was tested at an excitation frequency of 1Hz using an applied heating ramp of 3°C/min from 25°C to 230°C.

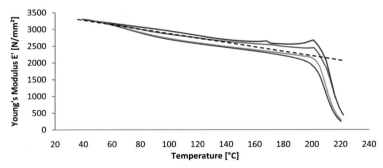

Figure 20 Storage modulus dependency on the temperature (four different measurements), measured using DTMA setup

Fig. 20 shows change of the storage modulus against temperature. It can be observed that the Young's modulus changes nearly linearly up to a temperature of 180°C. Similar results can be found in the literature [4, 51]. Above 180°C, the glass transition temperature is reached and the storage modulus is decreasing. The measured glass transition temperature of the DTMA results averages at about 210°C. The average Young's modulus at room temperature (3248N/mm²) decreases at a temperature of 180°C to 2399N/mm². The following equation is using a linear approximation of the temperature dependency of the Young's modulus and can be applied for a cured material below the glass transition temperature

$$E_{(T)} = E_{(0°C)} - a_E \cdot T = 2873 - 4.91 \cdot T \; [°C] \,. \tag{3.22}$$

Similar to the investigation of the thermal dependency on the Young's modulus, the CTE thermal behaviour can be observed. For representation of the right thermally induced stresses, the CTE of the resin is a sensitive parameter. Knowledge about the type of temperature dependency and the possibility of linear or non-linear approximation is important and has been observed using a thermo mechanical analyser (TMA / TA instruments). A squared cured resin sample with dimensions 10mm*10mm*2mm was placed inside the expansion setup and heated up at a ramp of 3°C/min from 40°C to 220°C.

In Fig. 21 the measured CTE for cured RTM6 is shown. First of all, it is visible that the CTE depends on the temperature. The CTE at room temperature is $53.5 \cdot 10^{-6}/°K$, at 120°C $70.1 \cdot 10^{-6}/°K$, and at 180°C $78.4 \cdot 10^{-6}/°K$. This indicates that a linear approximation does provide reliable results up to a temperature of 180°C. It was observed that by reaching the glass transition temperature the CTE increases dramatically.

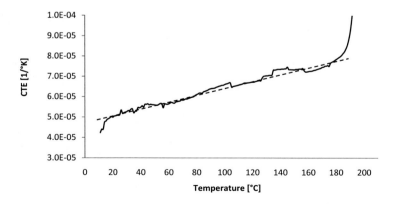

Figure 21 CTE dependency on the temperature, measured using TMA setup

In the literature, published by Johnston et al. [4] and Hobbiebrunken [51] linear approximations of the temperature dependency for CTE are used. This might be sufficient for totally cured resin (degree of cure almost one) at lower temperatures (range from RT to 170°C). The following relationship is proposed (dashed line in Fig. 21)

$$\alpha_{M(T)} = \alpha_{M(20°C)} + a_{MA} \cdot T = 5.0 \cdot 10^{-5} + 8.3 \cdot 10^{-8} \cdot T \; [°C]. \tag{3.23}$$

Thermo mechanical behaviour of the resin
The stress-strain behaviour of resin dependent on different temperatures has been studied by Hobbiebrunken [51] experimentally. He performed tensile tests at different temperature levels: room temperature, 120°C, 150°C and 180°C. Similar experiments have been performed in this work using sample geometry of 10*3mm dimension conforming to the DIN EN ISO 527-2. The tensile test was performed using a Zwick 1474 with an applied crosshead speed of 3mm/min.

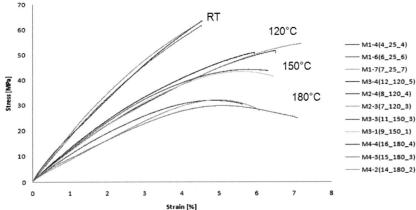

Figure 22 Stress – strain curves measured at different temperatures (25°C, 120°C, 150°C,180°C)

The experimental stress-strain curves show that the Young's modulus, the yield point and fracture depend strongly on the temperature. The resin behaves in a more ductile manner at higher temperatures. The yield stress (0.05% offset) drops from 43.9MPa at room temperature (RT) to 24MPa at the regular curing temperature of 180°C. The fracture stress behaves similarly, its value changes from 67.8MPa at RT to 27.6MPa at 180°C. In comparison to the DMA experiments, a deviation of the module has occurred, which can be explained by the used test method. Normally, DMS strain gauges are used to measure the strain during the experiment. Due to the high temperatures, this was not possible for all experiments. Therefore, the crosshead displacement was used to evaluate the strain.

Therefore the absolute evaluated values are questionable but the physical representation can be analysed. The following table 3.7 summarises the evaluated parameters.

Table 3.7 Average values

Parameters	25°C	120°C	150°C	180°C
Young's Modulus [MPa]	1949.87	1301.23	1239.6	1035.63
Standard deviation	119.95	53.5	64.69	81.89
Yield stress [MPa]	41.74	29.23	33.17	24.32
Standard deviation	2.96	3.37	1.40	3.09
Fracture stress [MPa]	60.04	51.61	44.02	27.63
Standard deviation	3.39	2.26	3.01	2.65

To take into account the non-linear material behaviour a temperature dependent yield criterion and fracture criterion have to be defined. This yield criterion can base on an excising yield criterion like Von Mises, Tresca, Drucker Prager, etc. In this work the Von Mises criterion will be used which is applicable to ductile materials. The RTM6 resin, however, shows brittle failure behaviour at RT and a ductile behaviour at higher temperatures. Hence, the Von Mises criterion might be sufficient, but can lead to a wrong estimation of the yield stress. The Von Mises stress is defined as follows

$$\sigma_M = \sqrt{\sigma_x^2 + \sigma_y^2 + \sigma_z^2 - \sigma_x\sigma_y - \sigma_x\sigma_z - \sigma_z\sigma_y + 3\left(\tau_{xy}^2 + \tau_{xz}^2 + \tau_{yz}^2\right)}. \qquad (3.24)$$

Using a linear shift factor (eq. 3.25 and 3.26) a temperature dependent yield and fracture criterion was applied. In comparison to the average values of yield and fracture stress a linear dependency was found:

$$\sigma_{Y_{(T)}} = 43.94\, MPa - 0.099 \cdot T \ [°C] \quad , \qquad (3.25)$$

$$\sigma_{F_{(T)}} = 67.75 MPa - 0.18 \cdot T \ [°C] \quad . \qquad (3.26)$$

The two criteria were used to define regions where yielding or degradation occurs:

$$\sigma_{(\varepsilon,T)} = \begin{cases} E_{(T)}\varepsilon^{el}, & \text{if} \quad \sigma \ < \sigma_{Y_{(T)}} < \sigma_{F_{(T)}} \\ \sigma_Y + \beta\varepsilon^{pl}, & \text{if} \quad \sigma_{Y_{(T)}} < \sigma \ < \sigma_{F_{(T)}} \\ \approx 0 \ , & \text{if} \quad \sigma_{Y_{(T)}} < \sigma_{F_{(T)}} < \sigma \end{cases} \qquad (3.27)$$

In the case of a stress state lower than the yield stress, the material behaves linearly. In the case of a stress state which is higher than the yield stress, but lower than the fracture stress, the material behaves by a bilinear elastic-plastic approach $\sigma_Y + \beta\varepsilon^{pl}$. In the case of a stress state which is higher than the fracture stress, degradation occurs and the stiffness will be set to nearly zero. This formulation is used as constitutive law to describe the plasticity and degradation.

3.8 Relaxation behaviour

A common method to represent viscoelastic materials is to use spring and damping elements. Simple standard forms of these representations are the Maxwell element, a combination of a spring and a damper element in series, and the Kelvin element, a combination of a spring and a damper element in parallel. In general, viscoelastic material behaviour can be classified into linear and nonlinear viscoelastic approaches, dependent on whether the load case has an influence on the relaxation behaviour. Another classification can be found by the temperature dependency. It can be classified into thermo rheological simple and complex materials. For application to polymerisation, thermo chemical approaches are available. In these models the characteristic relaxation is also dependent on the degree of cure.

One of the crucial points in the analysis of internal stress during the manufacturing process is the description of the relaxation effect. During the manufacturing process the relaxation will be dependent on time, temperature and degree of cure. There is no existing standardised measuring method to characterise the relaxation during curing, but for cured material there are. A common method uses forced oscillation. In this method different types of excitation like a mechanical excitation using Dynamical Mechanical Analysis (DMA), a temperature excitation using Differential Scanning Calorimetric (DSC), or an oscillating electrical field measuring the influence on the dielectric properties, is applied. The time shifted response in changed amplitude is used to derive the viscoelastic behaviour. A comparison between these measuring types can be found in Wenzel [22]. One of the most common methods to measure the viscoelastic properties is using mechanical exactions using DMA measuring device. For thermo rheological simple polymer materials this measuring method in connection to the Time Temperature Superposition (TTS) approach is widely conventional. In this case in the time (creep or relaxation tests) or frequency domain (test with different strain rates) test are done in a measuring window. The measured response curves are then condensed to a master curve which can be extrapolated to other time or temperature ranges.

In the following, a Dynamical Mechanical Thermal Analysis (DTMA) will be used with the Time Temperature Superposition (TTS) approach to evaluate, if a definition like thermo rheological simple can be applied to describe the relaxation effect during the manufacturing process.

To measure the relaxation a three-point bending test setup with a fully cured resin square plate with dimensions 35mm*10mm*2mm was tested at isothermal temperatures in the range of 40°C to 200°C with increments of 20°C. In each temperature increment an initial displacement of 0.1% strain was applied and the stress relaxation was recorded for 60min. Fig. 23 shows the relaxation profiles of the fully cured resin sample over total time. On the left axis the relaxation modulus (red curves) is shown on the right axis the corresponding temperature (blue curves).

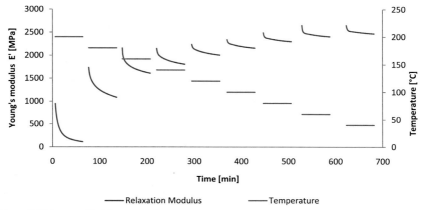

Figure 23 Relaxation Modulus over total time

Analysing the relaxation profiles over decay time, one important effect is visible. The relaxation profiles are almost linear shifted in the temperature range from 40°C-160°C. The relaxation profiles 180°C and 200°C show a faster relaxation. This effect can be explained by the impact of the glass transition temperature which was already measured in the previous study (Figure 20) because the modulus changes highly in the proximity of the glass transition temperature. In figure 24 the measured results are presented over decay time (relaxation time) and temperature in 3 dimensional way.

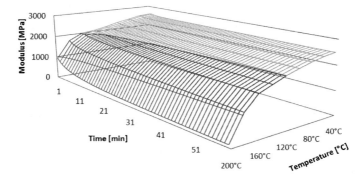

Figure 24 Relaxation behaviour over 1h dependent on different temperatures

Using the TTS approach, a shift factor a_T, a reduced time ξ and a master curve can be obtained. The master curve can be modelled using continuous or discrete formulations.

Discrete formulation, like a Prony series, is commonly used for viscous elastic modelling because it provides a good accuracy, but the evaluation of the needed parameters is complex. Additionally, for the curing process the relaxation behaviour depends on the material state and, as explained, on the glass transition temperature therefore a continuous closed solution might be more effective because a dependency of the glass transition temperature on a parameter can be defined more easily. As a first step, closed formulation using a reduced time has to be used

$$\xi = \int_0^t \frac{1}{a_T} dt .$$
(3.28)

The time temperature shift factor a_T can be modelled using the Williams-Landel-Ferry (WLF) equation or with linear simplification. It has been shown that for cured resin a linear approach is sufficient but close to the glass transition temperature this is questionable (Figure 23)

$$a_T = -0.04983 \cdot T + 10.83 \ [°C]$$
(3.29)

Fig. 25 is showing the shifted master curve (reference temperature of 120°C) and the relaxation profiles.

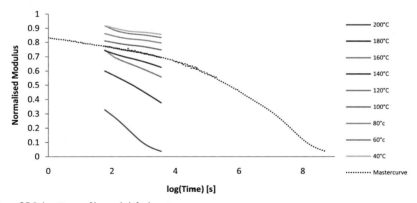

Figure 25 Relaxation profiles and shifted master curve

A modified stretched exponential formulation was used to model the relaxation behaviour. In this case, four parameters have to be evaluated, the relaxed modulus $E_{\infty(T)}$, the unrelaxed modulus $E_{a(T)}$, the characteristic relaxation time τ and a fitting parameter b. This continuous closed approach was used instead of a discrete approach to have the possibility to add, in further investigation, a dependency on degree of cure of the relaxation:

$$E_{(\xi)} = E_{a(T)} + \left(E_{\infty(T)} - E_{a(T)}\right)\left(1 - e^{\left(-\left(\frac{\xi}{\tau}\right)^{b}\right)}\right),$$ (3.30)

$$E_{a(T)} = 2890 - 4.4 \cdot T \, , E_{\infty(T)} = 2510 - 7 \cdot T, \tau = 411, b = 0.4 \; [MPa, {}^{\circ}C] \; (3.31)$$

During the evaluation of parameters, a good agreement in the temperature range of 40 to 160°C was found, which is displayed in Fig. 26.

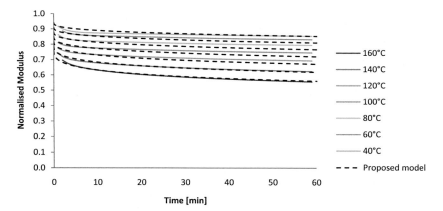

Figure 26 Model fit of relaxation profiles in the temperature range of 40°C to 160°C

Above the temperature of 160°C a good correlation was not found, because there is an influence of the glass transition temperature. This effect is changing the material behaviour and a linear temperature shift factor is sufficient anymore and a more complex shift factor formulation is needed.

Another possible approach using the differential form of the relaxation written incrementally can be found by Zocher et. al [28] and Svanberg et al. [25]. Hence, to define the stress at the actual time step, a relaxation term can be added. The stress increment is a sum of the derivative of the stiffness multiplied with the total strain increment and a stress relaxation component. The strain increment $\Delta\varepsilon_{total}$ consists of a sum of mechanical, thermal and shrinkage strain

$$\Delta\sigma_{(t)} = \sigma^{R} + \Delta C \cdot \Delta\varepsilon_{total} \, .$$ (3.32)

The relaxation stress increment describes the decrease of the stress and consists of the following relation multiplied with the so called recursive relation $S_{(t)}$ which is a state variable from the previous time step

$$\sigma^R = S_{(t)} \cdot \left(1 - e^{\frac{-\Delta \xi}{\rho}} \right).$$

(3.33)

The definition of the recursive element of the actual time step is defined as follows

$$S_{(t+\Delta t)} = e^{\frac{-\Delta \xi}{\rho}} \cdot S_{(t)} + \frac{\varrho \cdot C \cdot \Delta \varepsilon_{total}}{\Delta \xi} \left(1 - e^{\frac{-\Delta \xi}{\rho}} \right).$$

(3.34)

Consequently, the full relaxation behaviour bases on three parameters: ϱ, ρ and the reduced time ξ. The developed model is based on the description of the relaxation on cured resin. Furthermore, following the relation describes the resin behaviour dependency on temperature and glass transition temperature

$$\varrho = 0.0544 \cdot e^{0.0448 \cdot T} \ [°C],$$

(3.35)

$$\rho = 35.638 \cdot e^{0.0134 \cdot T} \ [°C].$$

(3.36)

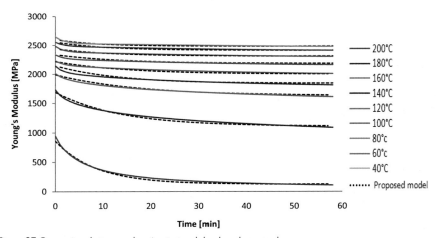

Figure 27 Comparison between relaxation test and developed approach

Using this formulation, a good approximation of the viscoelastic behaviour can be found from room temperature to 200°C for the cured material. The experimental measurement shows an increase of the relaxation behaviour, close to the glass transition temperature (for cured material around 202°C, Fig 27). During the curing process the glass transition changes dependent on the degree of cure. Hence, a dependency of the relaxation dependent on the degree of cure is needed. There are existing approaches which implement a cure dependency by adding the dependency to the shift factor of the TTS approach. The measurement of the stress relaxation in the rubber phase is not easy to undertake. The changes in view of curing are fast, so that stable conditions which are needed for the measurement cannot be reached. The measurement of the cured material has shown that close to the glass transition temperature, the relaxation is much faster. In the literature, Ruiz et al. [52] performed measurements of partial cured specimens. They found that the influence of the polymerisation degree is less and can be neglected. Researchers who have extended their models to implement a cure dependency are Paratasaya et al. [27] and Kim / White [26]. Only Kim / White shows a comparison of their model to experimental data. They also performed an experiment with partial cured specimens and measured the relaxation at RT.

To conclude, most of the researchers have observed the relaxation behaviour in the solid state, but not in the rubber state. A real measurement of the relaxation during curing can be found by Wenzel [53]. He compared different types of excitation like a mechanical excitation using Dynamical Mechanical Analysis (DMA), a temperature excitation using Differential Scanning Calorimetric (DSC) and an oscillating electrical field measuring the influence on the dielectric properties during isothermal curing in the rubber phase. He derived a relaxation spectrum and found a good description using a Vogel-Fulcher equation. Especially the idea of using dielectric spectroscopy for the characterisation of the relaxation could be promising, but it has to be remembered that relaxation during the rubber phase is very fast. Consequently, it is questionable if a detailed description dependent on temperature or degree of cure is really needed. In view of the measurement problem to derive a relaxation spectrum for the rubber phase and the level of complexity of the material model, a solution could be to find simple assumptions for the relaxation in the rubber phase. Following assumption will be supposed and used in this thesis:

1. The relaxation behaviour at the vitrification point is not dependent on temperature

2. The relaxation behaviour in the rubber phase is similar to the behaviour at 200°C

Using this simplification, a formulation can be used for the cured material and for the rubbery phase by setting the temperature of equation 3.28-3.32 to the maximal glass transition temperature in the viscous state.

4 Material characterisation – Composite

In the previous chapter the material is characterised for a single component. In a composite material two or more materials will come together and the behaviour will be a compound of the single properties. In the case of the manufacturing process, some of the single properties like resin module are not constant and dependent on time, degree of cure and temperature. In the following chapter, different homogenisation methods will be discussed to compute the process dependent homogenised material properties of the composite in a transient way. During the manufacturing process cure kinetic (exothermic heat flux), thermo dynamic (heat capacity, thermal conductivity, density) and mechanical parameters (engineering constants, thermal expansion coefficients, chemical shrinkage coefficients, characteristic relaxation times) have to be homogenised.

The evaluation of the mechanical behaviour of a cured composite material starts, in general, by taking the single fibre and matrix properties. These properties are used by micro mechanical approaches to compute the homogenised properties of the composite ply on the macro level. The micromechanical approaches are mostly analytical rules of mixture based on the appliance of the Voigt model in fibre direction and the Reuss model transverse to the fibre direction [54]. The Voigt model is derived from the examination that the fibre and the matrix can be treated as a parallel connection of springs. The Reuss model applies a serial connection of the fibre and matrix stiffness.

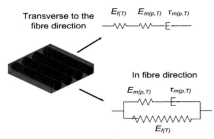

Figure 28 Schematic of material model

Two aspects have to be named. Firstly, during the manufacturing of composite the fibre volume content is changing due to thermal expansion and chemical shrinkage. Secondly, some parameters are not constant and behave dependent on time, temperature and degree of cure. To take this into account the homogenised properties needs to be recalculated during the transient analysis at each time step. Two general approaches can be applied; using numerical unit cell approaches or using analytical micromechanical equations. In the case of numerical unit cell methods it will lead to a two scale analysis which is computationally expensive. Therefore, the analytical approaches are preferred because they are fast and easy to implement. In the next chapters, different analytical homogenisation methods will be introduced and discussed on their abilities. The question will be if they are suitable, if resin parameters are not constant.

4.1 Heat of reaction

The volume heat flow coming from the reaction will be computed using the following equation

$$\dot{q} = \rho_{m(p,T)} H_{tot} \frac{dp}{dt}.$$
(4.1)

This equation can be extended by adding the fibre volume content to apply it for the homogenised composite

$$\dot{q} = (1 - V_f) \rho_{m(p,T)} H_{tot} \frac{dp}{dt}.$$
(4.2)

4.2 Density

The density of the composite is a first order tensor. The following simple approach can be found in the literature (Schürmann [54], Johnston [4]) in order to homogenise the density of the composite

$$\rho_c = \rho_f V_f + (1 - V_f)\rho_{m(p,T)} .$$
(4.3)

In case of a homogenous laminate with similar plies the laminate density is equal to the ply density $\hat{\rho}_c = \rho_c$.

4.3 Heat capacity

The heat capacity is a scalar quantity and the type of fabric and the laminate layup will have no influence on this value. In the Literature only one approach was found (Schürmann [54]), which is also used by Johnston

$$c_c = \frac{c_f \rho_f V_f + (1 - V_f)c_{m(p,T)}\rho_{m(p,T)}}{\rho_f V_f + (1 - V_f)\rho_{m(p,T)}}.$$
(4.4)

An experimental measurement of the specific heat capacity of a cured composite (quasi isotropic laminate, measurement in thickness direction), cured resin and a dry perform composite (quasi isotropic laminate, measurement in thickness direction) was done using Laser Flash Analysis (LFA) method in the temperature interval of RT to 200°C with discrete temperature steps of 25°C. The results of the process temperature range are presented in the figure below.

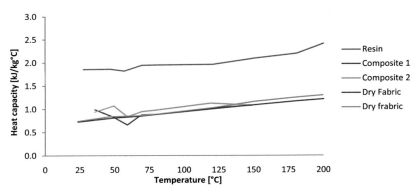

Figure 29 Measured heat capacity over temperature for cured resin and composite and dry fabric

In case of a homogenous laminate with similar plies the laminate heat capacity is equal to the ply heat capacity $\hat{c}_c = c_c$. In case of a laminate with miscellaneous plies following approach have to be applied

$$\hat{c}_c = \frac{\sum_{k=1}^{n} \rho_c \cdot c_c \cdot t_k}{\sum_{k=1}^{n} c_c \cdot t_k}. \tag{4.5}$$

4.4 Thermal conductivity

The thermal conductivity of a composite is a second order tensor which is strongly dependent on the type of fibre arrangement (fabric / unidirectional). There are existing analytical rules of mixture for unidirectional fibre arrangements which are based on the theory of parallel adjustment of the thermal conductivity in the fibre direction and a series connection in transverse direction of the fibre.

For a unidirectional layer an orthotropic/transverse isotropic behaviour of the thermal conductivity can be assumed (Schürmann [54]). There are also some extended versions of this simple model which are published by Kulkarni et al. [55], Springer and Tsai [56]:

Schürmann [54]

$$k_{c1} = k_f \cdot V_f + (1 - V_f)k_{m(p.T)},$$

$$k_{c2} = k_{c3} = \frac{-k_{m(p.T)} k_f}{k_f(V_f - 1) - k_{m(p.T)}V_f}. \tag{4.6}$$

Springer Tsai [56]

$$k_{c1} = k_f \cdot V_f + (1 - V_f)k_{m(p.T)},$$

$$k_{c2} = k_{c3} = k_{m(p.T)} \left(1 - 2\sqrt{\frac{V_f}{\pi}} + \frac{1}{B} \left(\pi - \frac{4}{\sqrt{1 - B^2 \frac{V_f}{\pi}}} \tan^{-1} \frac{\sqrt{1 - B^2 \frac{V_f}{\pi}}}{1 + B\sqrt{\frac{V_f}{\pi}}} \right) \right),$$

$$B = 2\left(\frac{k_{m(p.T)}}{k_f} - 1\right). \tag{4.7}$$

Kulkarni et al. [55]

$$k_{c1} = k_f \cdot V_f + (1 - V_f)k_{m(p.T)}, \tag{4.8}$$

$$k_{c2} = k_{c3} = k_{m(p.T)} \left(\frac{k_f(1 + V_f) + k_{m(p.T)}(1 - V_f)}{k_f(1 - V_f) + k_{m(p.T)}(1 + V_f)} \right). \tag{4.9}$$

A comparison between the different theories is performed in Fig. 30 for the thermal conductivity in transverse direction for totally cured resin and carbon HT fibres at room temperature $(k_{f\perp} = 1.7 \ W/m°K$) [54]. In the comparison the difference between Springer / Tsai and Kulkarni is small in contrast to the simple approach published by Schürmann.

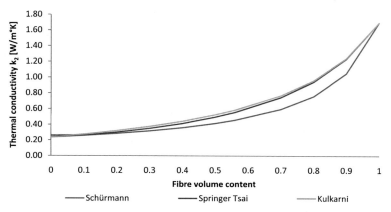

Figure 30 Comparison between different analytical homogenisation approaches for thermal conductivity

In the case of textile fabric the type of weave will influence the thermal conductivity. The used fabric (G1157) is of type plain weave with 96% of carbon fibres in warp direction and 2% of glass fibres in weft direction. Therefore, a simple analytical model will not give reliable results in transverse and out of plane direction. The idealisation of transverse isot-

ropy is not valid. On this account a voxel (volumetric pixel) based unit cell model was analysed using the software GeoDict. In Fig.31 the unit cell is displayed. (red=carbon fibres,
blue=glass fibres, green=binder, lucent=resin). In Table 4.1 a comparison between the
analytical approaches and the representative meso model of the fabric is presented using
the thermal conductivity values for totally cured resin and carbon HT fibres at room temperature. The fibre volume content is 60%.

Table4.1	Schürmann	Springer/Tsai	Kulkarni	GeoDict
k_{c1}	2.85	2.85	2.85	2.86
k_{c2}	0.48	0.56	0.58	0.99
k_{c3}	0.48	0.56	0.58	0.50

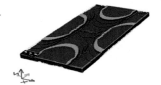

Figure 31 Representative Volume Element (RVE) of the textile G1157

An experimental measurement of the thermal conductivity has been performed of a cured
composite (quasi isotropic laminate, measurement in thickness direction) and a dry perform composite (quasi isotropic laminate, measurement in thickness direction) using Laser
Flash Analysis (LFA) method in the temperature interval of RT to 200°C. The results of the
process temperature range are presented in the Fig. 32 below.

The experiment comparison between a cured composite and a dry fabric show a significant change of the thermal conductivity by a factor of 7. In the RTM process, during the
first heating phase, before injection there will be only dry fabric inside the mould. Consequently these significant changes must be implemented into a material model [S2].

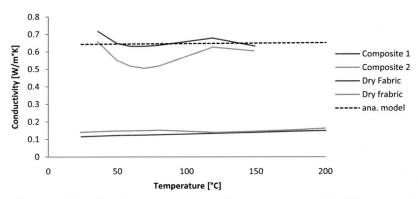

Figure 32 Measured thermal conductivity over temperature for cured composite and dry fabric in out of
plane direction

Similar to the observed resin behaviour the temperature dependency of the cured composite is small. In comparison to experimental achieved values the calculated conductivity in

out of plane direction is too low for the analytical based theories of Schürmann, Springer / Tsai and Kulkarni but also for the numerical achieved values from GeoDict. The deviation is forced by the value of the transverse conductivity of the fibre. The experimental determination of this value is difficult and in most applications the value is computed inverse. Accordingly for the given material system a transverse fibre conductivity of $k_{f\perp} = 2\,W/mK$ in application of the Kulkarni approach determines the transverse conductivity in a reliable way.

The thermal conductivity of a full laminate can be calculated analogue to the classical laminate theory [54]. Once the ply properties are defined for every ply that composes the laminate, the resulting value can be obtained by a conversion using polar transformation to the laminate coordinate system (ply orientation θ):

$$k_x = k_{c1|} \cdot \cos^2 \theta + k_{c2} \cdot \sin^2 \theta \ , \tag{4.10}$$

$$k_y = k_{c1|} \cdot \sin^2 \theta + k_{c2} \cdot \cos^2 \theta \ . \tag{4.11}$$

The in plane thermal conductivity $\hat{k}_x$ and $\hat{k}_y$ of the laminate is a parallel connection of the single ply values and accordingly the weighted sum of the laminate can be done to compute the resulting thermal conductivity:

$$\hat{k}_x = \sum_{k=1}^{n} k_{xk} \cdot \frac{t_k}{t}\,, \tag{4.12}$$

$$\hat{k}_y = \sum_{k=1}^{n} k_{yk} \cdot \frac{t_k}{t}\,. \tag{4.13}$$

In thickness direction a serial connection have to be applied. Consequently the value in thickness direction is much lower than in the in plane properties

$$\hat{k}_z = \frac{t}{\sum_{k=1}^{n} \frac{t_k}{k_{c3}}} \tag{4.14}$$

In case of a homogenous laminate with equal plies following simplification can be applied $\hat{k}_z = k_{c3}$

4.5 Engineering constants

The micromechanical approaches to homogenise the Young's modulus are mostly analytical rules of mixture based on the appliance of the Voigt model in fibre direction and the Reuss model transverse to the fibre direction. The Voigt model is derived from the examination that the fibre and the matrix can be treated as a parallel connection of springs. The Reuss model applies a serial connection of the fibre and matrix stiffness. Using these approaches, the following basic restrictions have to be fulfilled. The lamina has to be macroscopicly homogeneous, linear elastic and initially stress free. The fibres are homogeneous, linear elastic, isotropic, regularly spaced and perfectly aligned. The matrix has to be homogeneous, linearly elastic and isotropic [84]. It can be seen that restrictions such as being initially stress free for the lamina and linearly elastic behaviour for the resin are not fulfilled. Additionally, the fibre volume ratio changes due to thermal expansion and shrinkage. The magnitude will be around 2-3%. The solution to solve this dilemma might be to compute the homogenised properties incrementally. As expected, the composite transverse modulus is highly dependent on resin behaviour while fibre direction properties are much less affected.

In the literature, different micromechanical approaches for the transverse properties are available [54, 87] which are shown in Fig. 33.

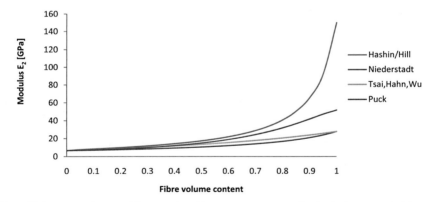

Figure 33 Comparison between different analytical homogenisation approaches for the transverse modulus

As shown in Fig. 33, the differences between the approaches are quite pronounced, especially for high fibre volume ratios. It must be mentioned that the maximum value for squared fibre packing is around 0.79 and for hexagonal packing is 0.90. By analysing the different approaches, it can be seen that the approach of Hashin / Hill [57] estimates the highest values for the transverse Young's modulus and are conservatively high. In this study, the homogenisation approach of Hashin / Hill was chosen, because a consistent approach for an orthotropic material exists (eq. 4.15-4.24). This approach is valid for a certain range of material but has to be proved for the materials used in this thesis later on. Hashin / Hill defined first the hydrostatic fibre K_f and matrix modulus K_m:

$$K_f = \frac{E_f}{2(1 - 2v_f)(1 + v_f)} \, ,$$
(4.15)

$$K_m = \frac{E_m}{2(1 - 2v_m)(1 + v_m)} \, .$$
(4.16)

They are used to determine the lateral bulk modulus K_l

$$K_l = K_m + \frac{V_f}{\left(\dfrac{1}{(K_f - K_m)} + \dfrac{1 - V_f}{(K_m + G_m)} \right)} \, .$$
(4.17)

For an orthotropic material, 9 elastic coefficients can be successively derived from the following equations:

$$E_1 = E_f V_f + E_m(1 - V_f) + \frac{4V_f(1 - V_f)(v_f - v_m)^2}{\left(\dfrac{V_f}{K_m} + \dfrac{1}{G_m} \dfrac{1 - V_f}{K_f} \right)} \, ,$$
(4.18)

$$v_{12} = v_f V_f + v_m(1 - V_f) + \frac{V_f(1 - V_f)(v_f - v_m)\left(\dfrac{1}{K_m} - \dfrac{1}{K_f} \right)}{\left(\dfrac{V_f}{K_m} + \dfrac{1}{G_m} + \dfrac{1 - V_f}{K_f} \right)} \, ,$$
(4.19)

$$G_{12} = G_m \frac{G_f(1 + V_f) + G_m(1 - V_f)}{G_f(1 - V_f) + G_m(1 + V_f)} \, ,$$
(4.20)

$$G_{23} = G_m + \frac{G_m V_f}{\dfrac{G_m}{(G_f - G_m)} + \dfrac{(K_m + 2G_m)(1 - V_f)}{(2K_m + 2G_m)}} \, ,$$
(4.21)

$$E_2 = \frac{2}{\left(\dfrac{1}{2K_l} + \dfrac{1}{2G_{23}} \dfrac{2v_{12}^2}{E_1} \right)} \, ,$$
(4.22)

$$v_{23} = \frac{E_2}{2G_{23}} - 1 \, .$$
(4.23)

In the case of transverse isotropy material, the following assumptions can be taken:

$$E_3 = E_2 \, , \qquad v_{13} = v_{12} \, , \qquad G_{13} = G_{12}$$
(4.24)

Experimental investigations of the Young's modulus in fibre and transverse to the fibre direction have been carried out using tensile experiments. Three different test panels are manufactured using RTM6 resin and 8 layers of CFRP unidirectional fabric G1157. The first panel was manufactured using the standard conditions at a curing temperature of 180°C (curing time 120min), a second test plate has been cured at 200°C (curing time 120min) and a third plate was cured at 120°C (curing time 360min). Tensile tests were performed in fibre and transverse directions. For each plate, 6 specimens are tested for the direction in fibre and transverse to the fibre direction. All tests predicated on the standard DIN EN 2587. The test has been performed by standard climate conditions with climate specimens at 23°C and 50% humidity (wet condition). In a separate experiment, the fibre volume content was determined according to DIN EN 2564 which was 59%. In the following table 4.2, the results are summarised and compared to the values of the official datasheet "75_t_2_0602_1_1" [29]. The following table 4.2 summarises the evaluated experimental data [S6]:

Table 4.2.

Results in fibre direction	$R_{\parallel}$ [N/mm^2]	Standard Dev. [N/mm^2]	E_1 [N/mm^2]	Standard Dev. [N/mm^2]
Curing temperature 200°C	1602.2	63.0	122920.4	2374.1
Curing temperature 180°C	1746.4	58.0	124504.0	3082.3
Curing temperature 120°C	1904.0	32.6	123355.9	4488.6
Datasheet G1157/RTM6(dry condition)	1970.0	130.0	136000.0	2000.0
Datasheet G1157/RTM6(hot/wet condition)	1850.0	120.0	136000.0	2000.0

Results transverse (90°) to the fibre direction	$R_{\perp}$ [N/mm^2]	Standard Dev. [N/mm^2]	E_2 [N/mm^2]	Standard Dev. [N/mm^2]
Curing temperature 200°C	34.7	1.61	9285.8	295.7
Curing temperature 180°C	33.1	4.37	8367.8	91.60
Curing temperature 120°C	39.9	2.10	8773.1	155.6
Datasheet G1157/RTM6(dry condition)	67.0	9.00	9400.0	200.0
Datasheet G1157/RTM6(hot/wet condition)	26.0	4.00	8500.0	200.0

Results in 45° direction	$R_{\parallel\perp}$ [N/mm^2]	Standard Dev. [N/mm^2]	G_{12} [N/mm^2]	Standard Dev. [N/mm^2]
Curing temperature 200°C	31.0	1.08	4683.1	36.3
Curing temperature 180°C	35.3	3.02	4702.5	64.8
Curing temperature 120°C	35.6	2.98	4399.7	72.5
Datasheet G1157/RTM6(dry condition)	62.0	5.00	5400.0	400
Datasheet G1157/RTM6(hot/wet condition)	49.0	4.00	4500.0	400

The experimental results show that stiffness and strength values are dependent on the curing conditions. The plate which is cured at a low temperature (120°C) has a higher Young's modulus in fibre direction and transverse direction. This leads to the conclusion that residual stresses are very high and initiate degradation by defects on the micro scale. There is also a large deviation of the strength values to the official datasheet values. In the datasheet there is no information about the curing conditions and manufacturing method. The deviation of the transverse strength is very high by a factor of nearly 2, compared to the dry condition values.

To compare the approach of Hashin / Hill [57] a comparison was performed on the average values of the experiments. The input values for Hashin / Hill are taken by Table 3.1 and 3.2.

Table 4.3

Experiment Average (cured at 180°)	Hashin/Hill ($V_f = 0.59$)	Difference
$E_1 = 124504 \, N/mm^2$	$E_1 = 118969 \, N/mm^2$	-4.4 %
$E_2 = 8367 N/mm^2$	$E_2 = 9135 \, N/mm^2$	9.1 %

The comparison shows especially for the transverse properties slightly too high values for the cured material at 180°C. In the next chapters 5 and 6.4 the reliability of the Hashin / Hill approach will be discussed in detail to explain the discovered effects related to the different curing conditions.

4.6 Thermal expansion

Similar to the engineering constant, the coefficients of thermal expansion (CTE) α_i of the homogenised ply can be defined using micro mechanical homogenisation. As a first question it is to define if there is a dependency of the matrix thermal expansion on the degree of cure. The proof of this is hardly difficult because there is no measurement method available to get the thermal expansion changes separated from the chemical induced shrinkage. From a literature review, the expansion coefficient was always defined with no dependency on the degree of cure. For the cured material different homogenisation approaches for the longitudinal and transverse behaviours can be found in the literature. All approaches for the longitudinal direction start from the requirement that fibre and matrix perform the same strain, are stress-free and influence of the transverse contraction are neglected. Again, some conditions are not or cannot be fulfilled. In the case of the engineering constants it was requested to have a strong influence of the resin modulus on the homogenised properties. For the thermal expansion coefficient in the fibre direction, this is also needed. For the cured material the assumption is valid that the stiff fibres compel the deformation of the matrix. During the process, first the thermal expansion coefficient α_1 is negative, after passing the gel point the coefficient starts to increase in dependency to the degree of cure. This dependency can be introduced if the equation for the homogenised properties involves the resin modulus.

In the transverse direction, the thermal expansion of the compound is matrix dominated. The theoretical fundamental is based on a connection in series. Using the assumption that the CTE of the matrix is not cure-dependent, the micromechanical model can be defined in the simplest way. In the following transverse and longitudinal approaches, from Schürmann [54], are chosen:

$$\alpha_1 = \frac{\alpha_m E_m (1 - V_f) + \alpha_f V_f E_f}{E_m (1 - V_f) + E_f V_f},$$ (4.25)

$$\alpha_2 = \alpha_m (1 - V_f) + \alpha_f V_f,$$ (4.26)

$$\alpha_3 = \alpha_2.$$ (4.27)

The approach of Schürmann was chosen because it is one of the most used and the restraint of the transverse strain can be neglected for low resin modules. As it can be seen in Fig.34, the influence of the resin modules for the CTE1 in fibre direction is realistic. For the transverse direction a linear dependency on V_f is assumed.

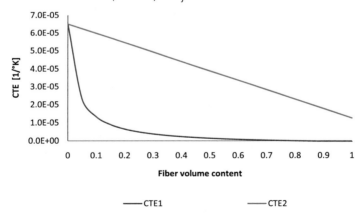

Figure 34 Comparison between the thermal expansion in fibre and transverse direction

The investigation about the thermal expansion behaviour of the resin (it was found that the thermal expansion coefficient depends on temperature, Figure 21) the composite should be analysed. For representation of the correct thermally induced stresses the CTE of the composite is a sensitive parameter. Knowledge about the type of temperature dependency and the possibility of linear or nonlinear approximation is important and has been observed using a thermo mechanical analyser (TMA / TA instruments). A squared cured composite sample with dimensions 7mm*7mm*1.6mm was placed inside the compression setup and heated up at a ramp of 3°C/min from 0°C to 180°C. This was done in fibre direction, in transverse direction and in out of plane direction.

In Figure 35 the measured CTE1 for cured composite in fibre direction is shown. First of all, it is visible that the CTE1 depends not strong on the temperature. The measurement of the CTE1 is afflicted with a high degree of uncertainty because the induced deformations are very small and therefore close to the precision threshold of the measurement system. Due to this reason the variability between the different measurements is very high. The overall precision of the TMA is given by $\pm 0.1\%$ of the sample size which leads to a threshold value $\Delta l = 0.007mm$. In the given application, if a $\alpha_1 = -0.4 \cdot 10^{-6}/°K$, a $\Delta T = 180°K$ and a sample size of $l_0 = 7mm$ is assumed, the resulting thermal contraction in fibre direction is $\Delta l = 0.000504mm$. Consequently the application of the TMA method is very questionable and not reliable. About the precision and reproducibility of measuring CTE using TMA method a statement can be found in the norm ISO 11359-2 [88]. In the annex A of this norm the results of a round-robin test of eight laboratories are shown for 8mm long specimens and a temperature increment of $\Delta T = 100°K$. For the range of a CTE from $5 \cdot 10^{-6}/°K$ to $1 \cdot 10^{-6}/°K$ a precision of $\pm 12\%$ and a maximum deviation of 61% is presented. For CTE values below $1 \cdot 10^{-6}/°K$ no values are available but precision and reproducibility will be worst.

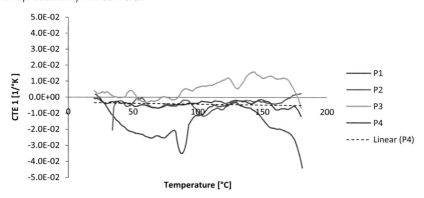

Figure 35 Measured thermal expansion in fibre direction (CTE1) over temperature

The average CTE1 in fibre direction at room temperature is $\alpha_1 = -1.9 \cdot 10^{-8}/°K$, at 120°C $\alpha_1 = -0.6 \cdot 10^{-8}/°K$, and at 180°C $\alpha_1 = -4.7 \cdot 10^{-8}/°K$. This indicates a linear approximation and does provide reliable results up to a temperature of 180°C. It was observed that by reaching the glass transition temperature the CTE increases dramatically. The dependency of the temperature can be idealised with the following linear dependency

$$\alpha_{1(T)} = \alpha_{1(20°C)} + a_{1A} \cdot T = -0.3 \cdot 10^{-6} - 1.0 \cdot 10^{-11} \cdot T \ [°C] \ . \qquad (4.28)$$

The CTE2 in transverse direction to the fibre at room temperature is $\alpha_2 = 20.41 \cdot 10^{-6}/°K$, at 120°C $\alpha_2 = 24.26 \cdot 10^{-6}/°K$, and at 180°C $\alpha_2 = 29.16 \cdot 10^{-6}/°K$. This indicates that a linear approximation does provide reliable results up to a temperature of 180°C. It was observed that by reaching the glass transition temperature the CTE decreases significant as shown in Fig. 36.

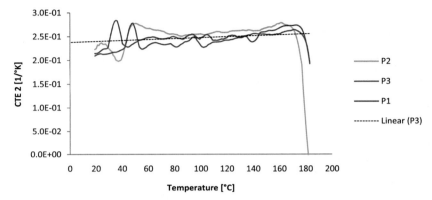

Figure 36 Measured thermal expansion in transverse direction (CTE2) over temperature

The dependency to the temperature can be idealised with the following linear dependency

$$\alpha_{2(T)} = \alpha_{2(20°C)} + a_{2A} \cdot T = 23.79 \cdot 10^{-6} + 1.0 \cdot 10^{-10} \cdot T \ [°C] . \qquad (4.29)$$

The CTE3 in thickness direction to the fibre at room temperature is $\alpha_3 = 4.8 \cdot 10^{-6}/°K$, at 120°C $\alpha_3 = 6.2 \cdot 10^{-6}/°K$, and at 180°C $\alpha_3 = 11.30 \cdot 10^{-6}/°K$. This indicates that also a linear approximation does provide reliable results up to a temperature of 180°C. In comparison to CTE2 in transverse direction it is visible that for this type of composite a transverse isotropic behaviour might not give reliable results (Fig. 37) because the CTE2 is four times higher than CTE3. Especially if a "spring-in" analysis is performed a simplification of the material to transverse isotropic behaviour will not give dependable results.

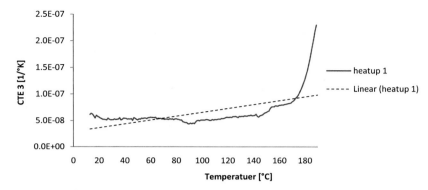

Figure 37 Measured thermal expansion in out of plane (CTE3) direction over temperature

$$\alpha_{3(T)} = \alpha_{3(20°C)} + a_{3A} \cdot T = 30 \cdot 10^{-8} + 4.0 \cdot 10^{-10} \cdot T \ [°C] \ . \tag{4.31}$$

In most simulation methods, a temperature dependency is neglected. Therefore, in the following Table 4.4 the coefficients are listed for a temperature interval of 20°C to 160°C. Additionally the standard deviation and maximum deviation of the round-robin test [88] are listed.

Table 4.4 CTE for linear analysis at RT

$\alpha_1 = -0.3 \cdot 10^{-6}/°C$	St. Dev. $> \pm 12\%$ [88]	Max. deviation 61% [88]
$\alpha_2 = 23.8 \cdot 10^{-6}/°C$	St. Dev. $\pm 2.6\%$ [88]	Max. deviation 3.7% [88]
$\alpha_3 = 3 \cdot 10^{-6}/°C$	St. Dev. $\pm 12\%$ [88]	Max. deviation 7.8% [88]

In comparison of the CTE2 in transverse direction and CTE3 out of plane direction it can be found that the behaviour of the material is not transversal isotropic. In case of warpage of shell like structures this effect can be ignored, but in case of spring-in of curved angle section this effect will have a significant influence. In this thesis the main objective will be focused to warpage of shell like box structures, therefore in the next section the material will be simplified as transverse isotropic.

Related to the changes of the material in curing process, the following definition can be assumed.

$$p < p_{gel}$$

$$\alpha_1 = \alpha_{f1} \quad ,$$

$$\alpha_2 = \alpha_3 = 0 \quad . \tag{4.32}$$

Below the gel point the resin module is nearly zero. Therefore only the fibre properties are taken into account After reaching the point of gelation a first polymer network exists and, therefore, the thermal expansion coefficients are introduced based on the approach of Schürmann [54] by the following relation:

$$p \geq p_{gel}$$

$$\alpha_1 = \frac{\alpha_{m(T,p)} E_{m(T,p)} (1 - V_f) + \alpha_{f1} V_f E_f}{E_{m(T,p)} (1 - V_f) + E_f V_f} , \tag{4.33}$$

$$\alpha_2 = \alpha_{m(T,p)} (1 - V_f) + \alpha_{f2} V_f . \tag{4.34}$$

Reasonable on the application of this model to analyse warpage the following assumptions for transverse isotropic material is used

$$\alpha_3 = \alpha_2 . \tag{4.35}$$

The thermal induced strains are taken into account by the gradient of the temperature ΔT multiplied by the CTE vector $\boldsymbol{\alpha_i}$ (eq. 3.29)

$$\Delta \varepsilon_{th} = \alpha_i \Delta T . \tag{4.36}$$

In chapter 6.4 the reliability of the chosen approaches is discussed in detail for different curing conditions.

4.7 Chemical shrinkage

For the determination of the influence of chemical shrinkage, a chemically induced shrinkage coefficient (CSC) γ_i can be introduced. Micromechanical approaches are useful and can be applied in a similar way as the thermal expansion coefficient (CTE) $\boldsymbol{\alpha_i}$ to the constitutive equations. Related to the CTE is the approach of CSC, which must be capable of applying only chemically-induced strain if the gel point is passed. The stiff fibres dominate the behaviour, prior to reaching the gel point. This behaviour can be obtained if the resin modulus is included in the micromechanical model. Two different approaches were found in the literature to homogenise the CSC from the micro to ply level published by Johnston and Wijskamp. [5, 21]. Both use the same equation for the behaviour in the longitudinal direction

$$\gamma_1 = \frac{(1 - V_f) E_m \gamma_m}{V_f E_{f1} + (1 - V_f) E_m} . \tag{4.37}$$

They differ in the transverse behaviour. Following Fig. 38, shows a difference between the published approaches for the homogenisation of the CSC in transverse direction and simple approach by neglecting the fibre properties. The total chemical resin shrinkage γ_m of 5.5% was assumed for the example.

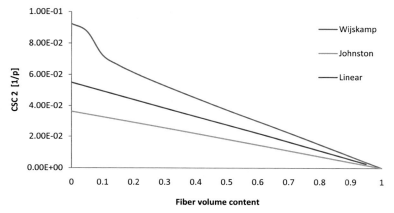

Figure 38 Comparison between different analytical homogenisation approaches for the chemical shrinkage transverse CSC2

The following homogenisation approaches can be found in the literature [4,24]

Wijskamp

$$\gamma_2 = (1 - V_f)(1 + v_m)\gamma_m + v_{12}\gamma_1 \ , \tag{4.38}$$

Johnston

$$\gamma_2 = (1 - V_f)(1 + v_m)\gamma_m - [v_{13}V_f + v_m(1 - V_f)]\left[\frac{\gamma_{m(1-V_f)E_m}}{V_f E_{f1} + (1 - V_f)E_m}\right] \tag{4.39}$$

Linear

$$\gamma_2 = (1 - V_f)\gamma_r \ . \tag{4.40}$$

For high fibre volume contents all approaches are show a similar behaviour. In case of a low ratio they start from different points. It is not evident why the published approaches by Johnston and Wijskamp are not starting from the total matrix shrinkage value of 5.5%. The measurement of the total amount of chemical induced shrinkage during a real process for example using Fibre Bragg Grating (FBG) sensors is not simple [S9]. The chemical shrinkage occurs after the gel point and is overlaid with the thermal contraction if isothermal condition, influenced for example by the exothermal reaction, cannot fulfill. It was tried to measure the chemical induced shrinkage using a FBG sensor system but common available sensors have a diameter of $200\mu m$ which is 25 times bigger than a carbon fibres. Consequently the difference in size was leading to no reliable results [S9].

Related to the changes of the material during the process, it can be assumed that during the liquid phase in the beginning of the curing process the degree of cure is below the gel point and no shrinkage occurs. Therefore the shrinkage CSC1 to 3 is set to zero. After reaching the rubbery phase between gel point and vitrification point the shrinkage is dependent on the stiffness of the resin. Therefore a stiffness dependent extension has been defined. If the solid phase is reached the shrinkage CSC is constant because the stiffness will not change large

$T \geq T_g$ and $p < p_{gel}$,

$$\gamma_1 = \gamma_3 = \gamma_2 = 0 \ . \tag{4.41}$$

After reaching the point of gelation a first polymer network exists and therefore the shrinkage coefficients are introduced

$p \geq p_{gel}$,

$$\gamma_1 = \frac{(1 - V_f)E_{m(T,p)}\gamma_m}{V_f E_{f1} + (1 - V_f)E_{m(T,p)}} \ , \tag{4.42}$$

$$\gamma_2 = (1 - V_f)(1 + v_m)\gamma_m -$$
$$\left[v_{13}\varphi + v_m(1 - V_f)\right]\left[\frac{\gamma_m(1-V_f)E_{m(T,p)}}{V_f E_{f1} + (1 - V_f)E_{m(T,p)}}\right]. \tag{4.43}$$

In case of transverse isotropy following assumptions can be found

$$\gamma_3 = \gamma_2 \ . \tag{4.44}$$

The chemical induced strain increment is based on the multiplication of the CSC γ_i and the increment of the degree of cure Δp

$$\Delta\varepsilon_{sh(p)} = \gamma_i \Delta p \ . \tag{4.45}$$

4.8 Relaxation behaviour

Similar to the previous experiments with resin samples in chapter 3 a dynamical mechanical thermal analysis (DTMA) is used to characterise the temperature dependent relaxation behaviour of the cured composite material. The experiments are performed transverse to the fibre direction, in 45° direction to the fibre and in fibre direction with a film fibre tensile test device. Totally cured dry G1157/RTM6 squared specimens have been tested in the range of 60°C to 200°C with isothermal temperature increments of 20°C. In each temperature increment an initial displacement of 0.1% strain was applied and the stress relaxation was recorded for 60min. Figure 39 shows the relaxation profiles of the fully cured samples over total time in fibre direction.

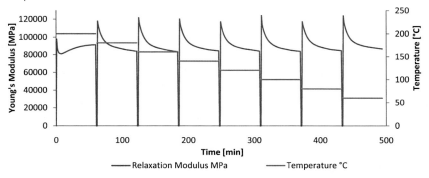

Figure 39 Measured relaxation module in fibre direction at different temperatures from 200°C to 60°C for 1h

The result of the performed experiment in fibre direction is that relaxation in fibre direction occurs, which was not expected. The used material is made of a unidirectional fabric. The fibres are not straight and they are undulated, in fact, of the weft yarn. On the meso level the stress in fibre direction and the undulation impose a stress component in out of plane direction. In this case, the resin and fibre are not connected in a parallel way. Relaxation occurs similarly to the transverse behaviour. For the given experiment, only a very thin (one layer, t=0.3mm) and short specimen (16mm) was tested. The length of a representative volume element (RVE) of the textile structure for G1157 is around 8mm. Therefore, the effect is amplified by these geometrical parameters. In fibre direction the relaxation over the temperature range is nearly constant and no large temperature effect can be observed. In summary, the measurement method is not reliable and these results will not be used. Additionally creep experiments have been performed to prove creep behaviour in fibre direction on coupon level in [S8]. In this case no creeping in fibre direction was visible.The following Figure 40 shows the relaxation behaviour in transverse direction.

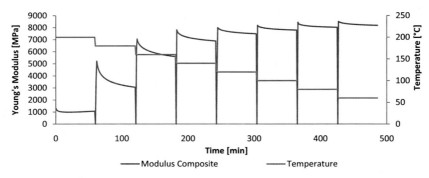

Figure 40 Measured relaxation module transverse to the fibre direction at different temperatures from 200°C to 60°C for 1h

The experiment transverse to the fibre direction shows the expected behaviour in comparison to the pure resin tests (Figure 23). The temperature effect is large and the influence of the glass transition temperature is visible by the drop of the modulus and a huge change in the relaxation behaviour. The following Figure 41 the relaxation behaviour in 45° to the fibre direction is shown.

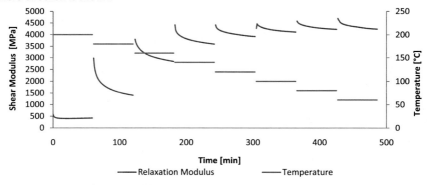

Figure 41 Measured relaxation module in 45° direction to the fibre at different temperatures from 200°C to 60°C for 1h

The behaviour of the material tested in 45° direction demonstrates a similar behaviour as the material tested in transverse direction. Again, the influence of the resin is dominating which is clearly visible by the influence of the temperature and the glass transition temperature. It has to be mentioned that the test in the 45° direction is not a pure shear test and the stress situation complex (no normal stress condition). Consequently this test results will not be used for the next comparison between model development and measurements.

The complete experimental determination of the viscoelastic properties for an orthotropic or transversal isotropic material is difficult because the time dependent Young's modulus, the shear modulus and Poisson's ratio have to be measured. The viscoelasticity is only based on the properties of the matrix. For a laminate, the fibre can be assumed to have a linear elastic behaviour. Meder [58] and Schürmann [54] have published the idea to introduce a quasi elastic solution for the viscoelastic description to describe the matrix behaviour and to use this time dependent matrix properties in the homogenisation method.

During a manufacturing process the fibre volume content can vary. The fibre volume content can change locally by the influence of chemical and thermal shrinkage but also globally by changes of the mould cavity, compaction and resin flow. Therefore a description of the viscoelastic behaviour based on the fibre volume content is needed. Meder [58] presented the following approach to homogenise the relaxation behaviour using homogenised relaxation factors r_i of the matrix component to the composite behaviour

$$r_1 = \frac{1}{1 + \dfrac{V_f}{(1 - V_f)\dfrac{E_m}{E_{f1}}}},$$ (4.46)

$$r_2 = \frac{1}{1 + \dfrac{V_f}{(1 - V_f)\dfrac{E_{f2}}{E_m}}}.$$ (4.47)

In the following diagrams, the relaxation of the transverse modulus for curing temperature of 180°C (left side) and room temperature (right side) is shown as three dimensional diagrams over the fibre volume content V_f.

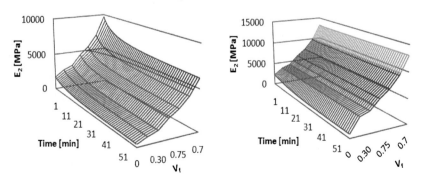

Figure 42 Relaxation spectrums transverse to the fibre direction dependent on the fibre volume content for 180°C (left side) and RT (right side)

The proposed approach shows a suitable solution over the full range of fibre volume content without any discontinuity. In the following Fig. 42 the relaxation behaviour in fibre direction is presented. Again, the left diagram shows the behaviour at curing temperature of 180°C and the right side at room temperature.

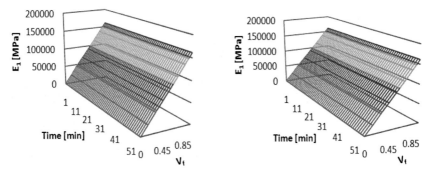

Figure 43 Relaxation spectrums in fibre direction dependent on the fibre volume content for 180°C (left side) and RT (right side)

As expected, behaviour in fibre direction s dominated by the elastic behaviour of the fibre, therefore, no relaxation occurs. The experimental measurements of the matrix (Fig. 22-23) are used and implemented into the rules of mixture (eq. 4.46 -4.47). The resulting Young's modulus is compared to the DMA measurements which were performed transverse to the fibre direction (Fig. 40).

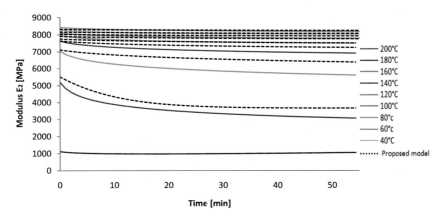

Figure 44 Comparison between measurements and development approach of Meder [58]

As a first solution, the approximation for this approach shows a large deviation and is not suitable to describe the behaviour, especially for high temperature in the near of the glass transition temperature. The given material is a fabric and it will be shown in chapter 5 that a linear dependency on the micro level does not exist because there are interactions between process micro degradation, plasticity and stress relaxation. Therefore, this influences the relaxation behaviour. A temperature dependent correction factor was introduced by the author

$$k_{com} = abs(1 - 0.0002 \cdot e^{0.0438 \cdot T}) \quad [°C].$$

(4.48)

Using this correction, the resin relaxation can be easily transferred to the composite behaviour without performing additional experiments. The following Figure 45 presents the accordance of model and measurement.

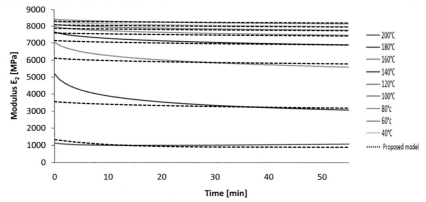

Figure 45 Comparison between measurement and development approach with correction factor

The error of the approximated solution at room temperature is lower than to the end of the experiment for high temperatures. This error can be accepted because the relevant time of relaxation is around 60min after curing. To achieve a more accurate solution, it is possible to use a Prony series to approximate module decrease. If Prony series are used, the number of material parameters increases significant because series of up to 16 terms have to be used [S1] and the computational procedure will be complex because the Prony terms have to be defined as state variables. For Prony series with 16 terms a state variable block of 729 numbers has to be used [21], which increases the computational effort dramatically. Therefore, the developed solution might be less accurate, but computationally effective for large problems. The development approach takes the relaxation behaviour for the cured material with respect to variation of the temperature from room temperature up to 200°C, into account.

The model takes into account a variation of the fibre volume content and is, therefore, suitable for manufacturing process in which a variation of the fibre volume content occurs.

5 Formation of residual stress on the micro level

Chapter 5 discusses different physical effects which were discovered in the characterisation chapters 3 and 4 and their influence on the formation of residual stress on the micro level. This detailed study was done to evaluate the sensitivity of different parameters to interpret their effects and to verify homogenised residual stresses on the macro level. In the following chapter, a numerical analysis of a composite unit cell is done to discuss the development of process-induced stress and the influence of the experimentally discovered effects from chapter 3-4.

In the last decades many researchers investigated the development of process-induced stresses on a macro level and also on the microscopic level [4, 21, 24, 54, 59]. The validation of process-induced stresses on the macro level is difficult. In fact, there are no nondestructive or destructive validated tests available to measure stress in composites. The development of process-induced deformations is mostly understood and has been modelled [4, 21, 24], but stresses are influenced by nonlinear effects like viscoelasticity, microscopic yielding and degradation, temperature and cure-dependent engineering constants. The question of choosing micro or macro approaches can be addressed to the physical effects which occur. Hobbiebrunken et al.[51] showed that thermoset resins behave thermo mechanically complex, effects like micro yielding and degradation dependent on the temperature affect the maximum microscopic stress level significantly. Therefore, an analysis of the formation of residual stresses on the micromechanical level can be an important tool to interpret discovered effects and verify homogenised residual stresses on the macro level.

In the literature, two general approaches are available on the issue of fibre and matrix distributions inside composites. The first approach assumes a regular geometric distribution of fibre and matrix by squared or hexagonal unit cells, the second considers larger unit cells of random fibre arrays. The first approach has the advantage that many researchers already have developed analytical solutions. For example Puck [60] has published a method to calculate the strain magnification of the matrix by applying external loads transverse to the fibre direction. These approaches might not represent the real topology of the structure but using geometrical simplifications, such as a unit cell, efficient analytical micromechanical equations can be derived. Therefore, a regular geometric distribution will be chosen in this study. This study attempts to investigate the effect of combined chemical and thermal strains using a coupled curing and mechanical simulation. Using this approach, the curing of the resin is simulated and linked to the development of the engineering constants, curing strains and changes in the thermal expansion coefficient. This chapter focuses on nonlinear effects in the resin. The results of the experiments of the previous chapter are used and integrated stepwise into a micro finite element model.

5.1 Model description

The objective of the present study is a micromechanical modelling of a fibre matrix unit cell. Typically, a fabric is made of yarns which consist of several thousand fibres and matrix unit cells. Packing models idealise the arrangement of fibres and matrix. Generally, two idealisations can be found using squared or hexagonal packing. This geometrical idealisation determines the theoretical maximum fibre volume content which is approximately 78% for squared packing and 90% for hexagonal packing. In Figure 46 a polished micro section of a composite is shown. It is obvious that a random distribution of fibre and matrix exists. In comparison of these types of packing it was found that a squared packing structure leads to higher thermal residual stresses of around 32%. Therefore, a squared idealisation of the unit cell was chosen for this study to take the extreme case.

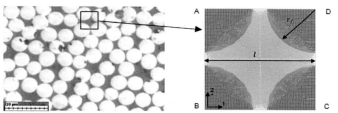

Figure 46 Polished micro section image of a CFRP composite   **Figure 47** FE mesh of the unit cell model

A parametric 2D model of the unit cell was built. The material of the polished micro section image was used for determination of the fibre volume content experimentally, using a DIN EN 2546 method. The fibre volume content was in average $v_f = 0.602$. Analysing the image (Figure 46) a fibre diameter of $d_f = 7\mu m$ was measured, leading to an effective length of the unit cell of $l = 8\mu m$:

$$l = \sqrt{\frac{d_f^{\,2} \cdot \pi}{4\, V_f}} \; . \tag{5.1}$$

Using this topology, a 2D plain strain model was built consisting of 13938 quadrangle elements (Figure 47). The boundary conditions are applied similar to the reference [51]. The left edge (A-B) is fixed in direction 1 and the lower edge (B-C) is fixed in direction 2. Movement of the right edge (C-D) is restricted to a uniform movement in direction 1 using multi point constraints. The upper edge (A-D) is constraint in the same way to move uniformly in direction 2. The reference material used for the polished cut image has the following curing history as shown in Figure 48.

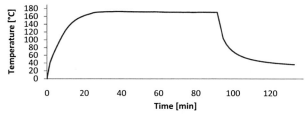

Figure 48 Temperature boundary conditions

During the process, the temperature was measured using a thermocouple. This temperature profile was applied as temperature boundary condition on the FE model.

5.2 Results of the process simulation

The results provided by the coupled thermo mechanical process analysis start, at first by
the development of the degree of cure and the development of the glass transition temperature during the process time.

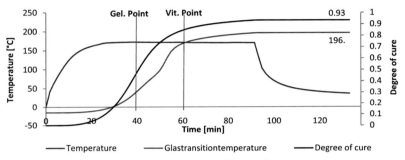

Figure 49 Simulation results of glass transition temperature and degree of cure vs. time

In the analysis, the cure kinetic approach of Kamal / Sourour is integrated [31]. Figure 49
shows the development of the degree of cure which reaches a value of 0.93 at the end of
the process. For the unit cell a uniform temperature was defined, therefore, only one result
curve is shown. Similar to the degree of cure the glass transition temperature is calculated
using the DiBenedetto equation with an extension published by Dykmann [36]. The development of the glass transition temperature is shown in Figure 49 with a final temperature
of 193°C. Using these curves two important process transition points can be identified: the
gel point and the vitrification point. As introduced in the previous chapter these two points
are used for identification of the state of the material between liquid and rubber as well as
rubber and solid.

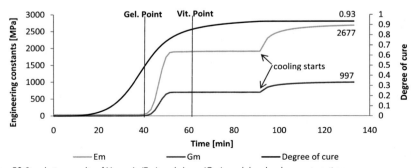

Figure 50 Simulation results of Young's (Em) and shear (Gm) modulus development vs. time

Figure 50 shows the development of the Young's modulus and the shear modulus during the curing process using eq. 3.20-21 coupled to the glass transition temperature. Similar to the experimental results shown Figure 18, a smooth development of the engineering properties in the rubber material state is defined. In addition to eq. 3.22 and the experimental results in Figure 20, the temperature dependency of the modulus is taken into account during cooling from the curing temperature to room temperature. This results in a Young's modulus of 2677MPa and a shear modulus of 997MPa. In the previous chapter yielding was discussed, but will be investigated later in more detail, as it leads to local variations of the modulus. Related to the dependency of the modulus the coefficient of thermal expansion (CTE) a chemical induced shrinkage coefficient (CSC) can be calculated and analysed using Figure 52. Applying eq. 3.23 the CTE has a value of $55.5 \cdot 10^{-6} 1/°K$. The dependency of the CTE on the degree of cure during the viscous material state and the dependency on the temperature during cooling down is clearly observable.

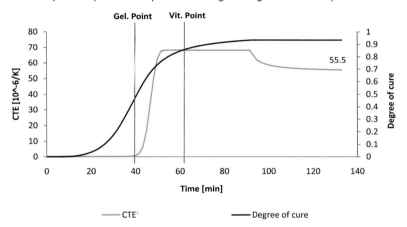

Figure 51 Simulation results of CTE development vs. time

The local properties of the matrix can be transferred to the composite using the global displacement of the unit cell. Figure 52 shows the scalar displacement in direction 1 measured at point C. The sum of the heat-up phase, the chemical induced shrinkage and the thermal induced shrinkage at cooling down lead to the resulting process-induced displacement. The contributions of the chemical and thermal induced shrinkages to the coefficient of thermal expansion (CTE) and chemical induced shrinkage coefficient (CSC) can be derived during the different phases of the curing process and compared to analytical micromechanical equations.

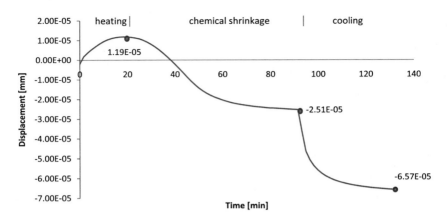

Figure 52 Simulation results of scalar displacement in direction 1 vs. time

For the CTE, the following analytical rule of mixture is available [54]:

$$\alpha_{22} = \varphi\alpha_{F_\perp} + (1 - \varphi)\alpha_M + \varphi(1 - \varphi)(\alpha_M - \alpha_{F_\parallel}) * \frac{E_{F_\parallel}v_M - E_M v_F}{\varphi E_{F_\parallel} + (1 - \varphi)E_M}$$

$$= 36.3 \cdot 10^{-6} 1/K,$$

$$\alpha_{22} = \frac{\Delta l}{l_0 * \Delta T} = \frac{(6.57 - 2.51) \cdot 10^{-5}}{8.0 \cdot 10^{-3} \cdot (170 - 36)} = 37.8 \cdot 10^{-6} 1/K. \tag{5.2}$$

The comparison of the results from the FE analysis shows a good agreement between the analytical approach and the unit cell results. Similar to the CTE, the CSC can be compared. For the CSC, different analytical micromechanical approaches are available as published by Johnston [4], Wijskamp [24]

$$\gamma_2 = (1 - V_f)\gamma_m = 0.0096 \ 1/p_{(93\%)} \ ,$$

$$\gamma_2 = \frac{A_{ges}}{A_{CSC}} = \frac{(1.19 \cdot 10^{-5} - (-2.51 \cdot 10^{-5}))}{0.008} = 0.0046 \frac{1}{p_{(93\%)}}. \tag{5.3}$$

The chemical shrinkage can be obtained from the volumetric shrinkage in association to the degree of cure p. For the given 2D application the percentage of shrinkage can be calculated using the total area of the unit cell during chemical shrinkage. The comparison shows a difference between the analytical (eq. 4.38) and numerical approaches of about 48%. This leads to the result that the analytical approach for the description of a homogenised chemically induced shrinkage coefficient (CSC) is not sufficient.

5.3 Study - Effects of matrix nonlinearities on the formation of residual stresses

In chapter 3 it was shown that the stress strain behaviour of the matrix is thermo mechanically complex (Fig. 22). Effects like temperature dependency on the Young's modulus and CTE, nonlinear temperature dependent stress-strain behaviour, micro yielding and viscoelastic relaxation effects will be discussed now by adding, stepwise, more complexity to the material law of the FE-model. Therefore, in this section, first a linear analysis is performed as reference. In this analysis, no temperature dependencies are taken into account. The material parameters are taken from Table 3.1- 3.2. As temperature load, the maximum process temperature (170°C) and the final temperature (36°C) are selected. In a second step, a transient analysis of the whole curing process is performed and the effect of cure shrinkage induced strain is added (eq. 3.19). The resin engineering properties are coupled to the degree of cure by eq. 3.20-21. In the third step, the temperature dependency of the Young's modulus and the CTE is added to step 2 by applying eq. 3.22-23. In a fourth analysis, the nonlinear stress strain dependency (eq. 3.24-27) is added to the model including micro yielding and micro degradation. In the fifth and last step the material model is extended by a relaxation term (eq.3.32-36). In the following table 5.1 the results maximum global displacement in direction 1, resulting global CTE, maximum local and global strain, maximum Von Mises and 1 principle stress component are listed for the 5 studies.

Table 5.1 Summary Study 1-5

	Study 1 Thermal shrinkage	Study 2 Thermal + chemical shrinkage	Study 3 Thermal + chemical shrinkage Temp. dependent properties	Study 4 Thermal + chemical shrinkage Temp. dependent properties Matrix yielding + degradation	Study 5 Thermal + chemical shrinkage Temp. dependent properties Matrix yielding + degradation Stress relaxation
max. global displacement	$28.2 \cdot 10^{-6}\ mm$	$68.9 \cdot 10^{-6} mm$	$72.4 \cdot 10^{-6} mm$	$72.5 \cdot 10^{-6} mm$	$7.17 \cdot 10^{-6} mm$
res. global CTE	$36.4 \cdot 10^{-6}/°K$	$33.9 \cdot 10^{-6}/°K$	$38.0 \cdot 10^{-6}/°K$	$38.5 \cdot 10^{-6}/°K$	$3.80 \cdot 10^{-6}/°K$
max. local strain	2.92%	7.2 %	7.4 %	7.7 %	7.3%
max. global strain	0.48%	0.90%	0.90%	0.90%	0.89%
max. local Von Mises stress	$64.4\ MPa$	$153\ MPa$	$134\ MPa$	$67\ MPa$	$67\ MPa$
max. local 1st principal	$38.8\ MPa$	$211\ MPa$	$182\ MPa$	$130\ MPa$	$72\ MPa$

Analysing the results given in Table 5.1 and Figure 53, the formation of residual stresses is given. In the first row the equivalent Von Mises stress is shown. This criterion neglects the influence of the hydrostatic stress state and assumes that yielding is independent. Therefore, the first principle stress is also shown. The Von Mises stress might be usable for thermoset resin because the thermal and chemical shrinkages lead to tensile stress components. It is visible that the sum of thermal and chemical shrinkages leads to a higher stress state with a value of 220MPa for the Von Mises stress (study 2). This value reaches the yielding and failure points and displays the need to extend the material law to the effect of temperature dependent material properties. Adding this effect (study 3), the maximum Von Mises stress decreases slightly to 178MPa, but the values are still very high and out of range of Hooke's law. Study 4 adds the nonlinear stress strain relation to the model (eq. 3.24-27). The results show a significant decrease in the residual stresses to 110MPa. The maximum of the stress changes from a point wise distribution to a more plane distribution. In the last step, stress relaxation is added to the model. The effect of relaxation will be faster at high temperatures than at room temperature; therefore, the effect of chemical-induced stresses on the level of stress is smaller. In comparison to the previous study, the distribution of the Von Mises stress is more uniform. Applying the information of temperature dependent yielding and failure, the regions where these effects appear can be displayed. Microscopic yielding occurs at the positions of high stresses. This yielding leads to a redistribution of the stress and additional yielding in regions where the matrix behaves linear (compare Study 3.) occurs. The redistribution decreases the total effort on the matrix but cannot avoid microscopic damage, which takes place close to the matrix fibre interface. The FE model was built to analyse the inner stress state of the matrix. The physical behaviour of the fibre matrix interface was assumed to be ideally aligned. Accordingly, at this point no differentiation can be found between matrix and interface failure.

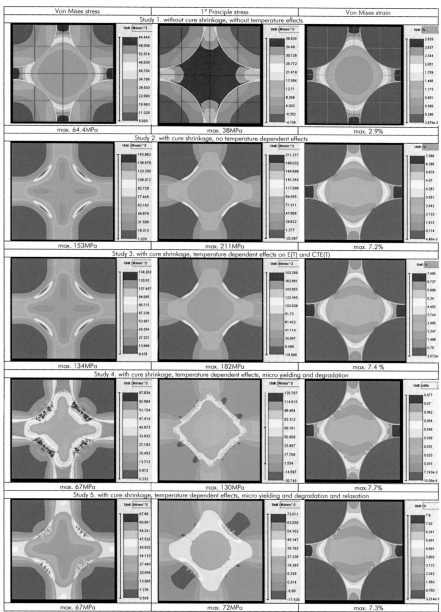

Figure 53 Results of the parameter study on the effect of matrix nonlinearities

71

The following Fig. 54 shows the regions of yielding and failure for Study 4 (without relaxation) and Study 5 (with relaxation). It can be observed that the description of the viscous elastic effect by stress relaxation is a sensitive parameter on the maximal amount of stress and process defects. The DMA experiment showed that the effect of relaxation on cured resin samples is important. Additionally, it was shown that a linear viscoelastic approach is not sufficiently close to the glass transition temperature. Taking a critical view on the numerical results it is obvious that the idealisation of linear viscoelasticity is insufficient. In Figure 54 the regions of yielding are displayed. Coming from the geometry of fibre and matrix an effect similar to notch stress appears. The increase will be affected in the way that a linear viscoelastic approach can be used but there will be an overestimation of stresses. This leads to the conclusion that the effect of viscoelasticity is not simple due to the two effects, softening close to the glass transition temperature and yielding / failure due to high micro stress values.

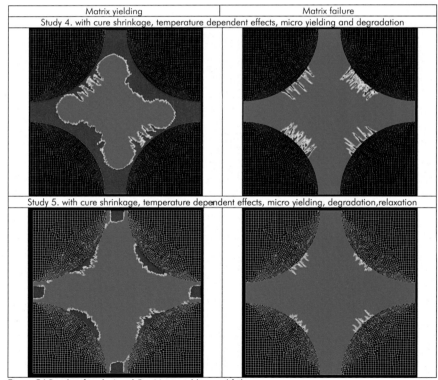

Figure 54 Results of study 4 and 5. - Matrix yielding and failure

5.4 Study - Effects of micro residual stress / defects on the macro level

The previous part demonstrated that combined chemical and thermal shrinkage during manufacturing can lead to high micro stress levels and to defects like micro yielding, micro degradation and fibre matrix interface failure. The sum of these effects will influence the resulting properties of the composite. In general, residual stresses increase strength and stiffness, because their nature is of type tensile and external loads have the opposite direction. Defects will decrease stiffness and strength properties, because they are the starting points of failure in case of external loading. Therefore, the transfer of the results on the micro level can be an important method to validate the stiffness, stress level and strength on the macro scale. The transfer between micro and macro scale is usually done by analytical or numerical homogenisation methods. One of the simplest homogenisation methods is to use the numerical model and to apply after the process simulation, a mechanical analysis of a tensile load. Using a displacement function the total reaction force can be calculated and macro stiffness and stress can be derived taking into account the geometrical dimension of the unit cell

$$\sigma_2 = \frac{\sum F}{A_{ges}}, \qquad E_2 = \frac{\sigma_2}{\varepsilon_2} \ . \qquad\qquad (5.4)$$

A linear displacement function is applied on side A-B (Figure 46) in negative direction 1 with the value of 1% macro strain. The side D-C is able to move during process simulation and fixed using DOF fixation in the tensile test. The total reaction force is used to calculate the strength and macro stress as the sum of forces taking node values of side D-C in direction 1. In the following Fig. 55 the stress-strain relation of the macro level is shown. It is visible that the macro response is not linear and a damage accumulation starts at a strain value of 0.8% and decreases the stiffness.

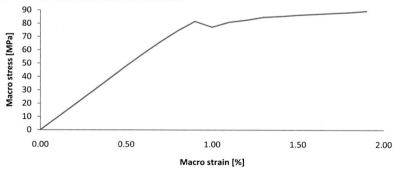

Figure 55 Macro stress-strain relation

The damage distribution can be shown using the damage variable (Figure 56) at discrete points. It is visible that the failure starts from the initial process defect / damage and grows taking a 45° angle from one fibre to another. In the case of residual stress-free configuration, a failure would occur between the upper and lower fibres first. In the case of this configuration, with process-induced stresses, the failure behaviour changes and the damage grows in the opposite direction.

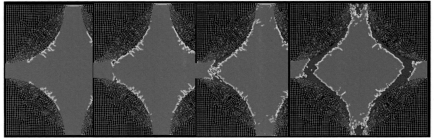

Figure 56 Development of matrix damage for macro strain value of 0.8, 1.1, 1.3 and 2.2%

The direction of the residual stress is shrinkage which leads to a tensile stress state. Consequently, the mechanical load has the opposite direction. Figure 57 shows this stress state schematically using black arrows for mechanical-induced stresses and red arrows for process-induced stresses.

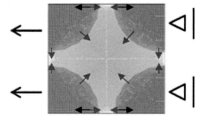

Figure 57 Stress state schematic with black arrows for mechanical-induced stress, red arrows for process-induced stresses

As one conclusion can be determined that high process-induced stresses change the failure behaviour on the micro scale, significantly.

In the following parameter study (Figure 58), the results of the analysis are compared to three different stress configurations without residual stresses and with three different residual stress configurations (thermal, thermal+chemical and thermal+chemical+non linear effects).

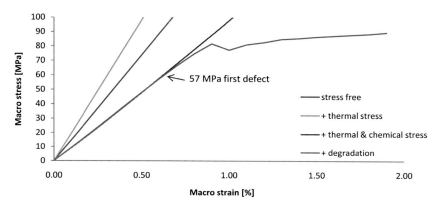

Figure 58 Influence of different residual stress states on macro stiffness

It can be observed that the residual stress states have a large influence on the stiffness values. In the case of residual stress free configuration the Young's modulus is **14733 MPa**. Adding only thermally induced stresses the stiffness increase and leads to a value of **19361 MPa**. In case of using the total sum from thermal and chemical induced shrinkage stresses the stiffness decreases to a value of **9539 MPa**. For the observed material (RTM6 / G1157) the following reference values can be taken from the datasheet [29] and from the performed tensile test (Chapter 4.5).

Table 5.2

Experiment	Process condition	E_2	Standard Dev.	$R_\perp$	Standard Dev.
	180 °C cured	8367 MPa	91 MPa	33.1 MPa	4.4 MPa
	120 °C cured	8773 MPa	155 MPa	39.9 MPa	2.1 MPa
Datasheet [29]					
hot / wet	unknown	8500 MPa	200 MPa	26 MPa	9 MPa
dry	unknown	9400 MPa	200 MPa	67 MPa	9 MPa
Simulation					
	180 °C cured	9539 MPa		57 MPa	

In comparison to these experimentally evaluated stiffness values and strength values, the numerical results show a good correlation. It has to be considered that fibre undulations of the fabric, the structure of the fabric are not taken into account in the FE - model and also the influence of moisture effects. Therefore, an overestimation of the stiffness shows the right physical behaviour.

Conclusion

In the presented study process-induced residual stresses of composite materials have been analysed using a process simulation of a squared unit cell including thermal and chemical induced shrinkage. Experiments show that the polymer matrix behaves nonlinear. Temperature dependent effects like yielding, fracture, relaxation have been added to the FE - model stepwise. The following results are found:

- Analysing process-dependent residual stresses only using temperature gradients is not sufficient and leads to wrong maximum stress values and wrong stress distributions,
- Adding chemical shrinkage to thermal shrinkage leads to unrealistically high stress values in the case of linear matrix behaviour. In reality, the polymer behaviour is thermo mechanically complex, and yielding and degradation occur. After matrix yielding, a redistribution of stress takes place, leading to a more uniform stress distribution,
- During the process, degradation and damage appear in the fibre matrix interface. This degrades the resulting composite strength and stiffness.

The process-dependent stress and defects have been used afterwards and studied to identify their effect on the macro scale. The following conclusions can be found:

- The failure behaviour changes in the case of a superposition with mechanical tensile load. The process-induced stresses change the points of maximum stress.
- The stiffness is influenced by the process-induced stress.

6 Formation of residual stress on the macro level

In chapter 6 the analysis method on the macro level will be developed and discussed in detail. First of all, a simulation strategy is formulated which should be applicable to large box structures. After this, a validation will be presented. This validation is performed on the coupon level using two test cases. First, a proof is done to validate resulting engineering properties and stresses on the tensile tests results of Chapter 4 and second, the correct determination of process-induced deformations using unsymmetrical laminates is compared.

6.1 Simulation strategy

As shown in chapter 2 about the material characterisation, many single quantities influence the formation of process-induced deformations and stresses. The manufacturing process of composite parts is a multi-physical problem, because it connects the thermodynamic discipline (thermal conduction in mould and part), chemistry (curing of the resin) and mechanical analysis (warpage). Additionally, this multi-physical problem is applied to a multi-scale problem, because single components like resin, fibre and fabric are combined and used to build large structures like a wing cover, fuselage section, etc. The aim of the analysis method should be to represent the correct physical behaviour on the part level with necessary involved parameters to perform parameter studies, sensitivity analysis and parameter variability analysis to increase part quality, decrease manufacturing costs and to design stable manufacturing processes. Therefore, a simulation strategy is needed which allows a detailed analysis with less computational effort.

Two different possibilities are available to solve a multi-physical problem between thermal and mechanical analysis. First, a strong coupling is feasible, where thermal and mechanical equations are solved simultaneously. The second method can be a sequential coupling where a thermal analysis is performed and temperature profiles are transferred as boundary conditions, namely, temperature loads, to the mechanical analysis. This sequential coupling can be done if the thermal analysis part is influencing the mechanical part, but not in the inverse direction. In the given case, the heat created by mechanical work can be negligible therefore, a sequential coupling is sufficient. This coupling enables different advantages because optimisation of temperature process parameters like heat rate, curing temperature and curing time in the thermal part can be done separately. On the mechanical side, parameter studies such as small variation of the laminate stacking, influence of draping errors, variation of material parameters can be analysed without recalculating the temperature loads. The sequential coupling offers the possibility to use different types of mesh and element types in the different analysis modules. For the given use case, the RTM process, the mould and part have to be analysed in the thermal analysis part. This has to be done using a volume mesh of the mould and part (Figure 59). In the mechanical part, the deformation and stress of the mould are not needed. Therefore, shell elements can be used. This enables to reduce the computational effort significantly and a variation of the layup can be done easily without changing the geometry.

Thermal Analysis   **Mechanical Analysis**

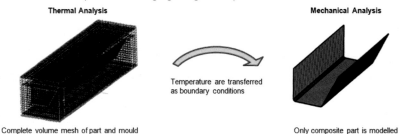

Temperature are transferred
as boundary conditions

Complete volume mesh of part and mould   Only composite part is modelled
                                                           using shell or volume elements

Figure 59 Simulation strategy

Another advantage of a sequential coupling will be that other manufacturing processes can be adapted. For example, the thermal module can be changed in the case of an autoclave curing process to a virtual autoclave. This virtual autoclave was developed in the research project Lokost / Probec (Luftfahrtforschungsprogramm IV) and consists of a semi-analytical approach to compute temperature boundary conditions dependent on autoclave parameters, such as fluid velocity, part position in the autoclave, part - part interaction in the autoclave etc.

To capture the multi-scale problem, the following method is recommended. During manufacturing of composites, the fibre volume content can be influenced and deviation from the "As-Planned" to the "As-Built" value appears. Dependent on the fibre volume content, all properties are changing. To take into account a possible variation of the fibre volume content and a variation of resin properties, the homogenised ply properties have to be recalculated during the transient analysis. Two general approaches can be applied by using numerical unit cell / representative volume elements approaches or using analytical homogenisation methods. Using numerical unit cell methods leads to a two scale analysis which is computationally wise, but more expensive. In consequence, the analytical approach is preferred. The use of analytical micromechanical homogenisation methods enables the following of the parameter changes. The advantage of this approach will be that the input parameters can be divided into resin and fibre properties. Experimental characterisation of single quantities is always easier as was demonstrated by chapter 3 and 4. Another advantage will be that the method provides additional results, because cured homogenised properties on the ply level are available. Therefore, the following material properties are provided: Young's moduli E_1, E_2, E_3, shear moduli G_{12}, G_{23}, G_{13}, Poission's ratios v_{12}, v_{13}, v_{23}, thermal expansion coefficients CTE1, CTE2, CTE3, heat capacity c_p, density ρ and thermal conductivities k_1, k_2, k_3.

6.2 Thermal analysis module

The necessity of performing a heat transfer analysis in the composite part is related to two different aspects. First, it depends on the thickness of the laminate. For thin laminates (thickness smaller than 2mm) the assumption of a uniform temperature in thickness direction is valid because exothermic reaction heat can be neglected. The second aspect is a temperature gradient within the part itself. Related to forced convection heating, the curing process is much faster in regions of higher temperature, compared to cooler regions. A rough rule-of-thumb states that increasing the temperature by 10°C doubles the speed of the curing reaction. To implement the exothermic heat flux from the resin cure reaction, a model developed by Bogetti and Gillespie [14] can be used. The governing equation of heat transfer is the transient anisotropic heat conduction equation, with a heat flux generation term from the exothermic cure reaction:

$$\rho_c \cdot c_{p,c} \frac{\partial T}{\partial t} = \nabla \cdot (\mathbf{k}_c \cdot \nabla \cdot T) + \dot{q}, \tag{6.1}$$

$$\dot{q} = (1 - V_f) \, \rho_{m(p,T)} \cdot H_{tot} \cdot \frac{dp}{dt}. \tag{6.2}$$

The influence of temperature and degree of cure on heat capacity, thermal conductivity and density is taken into account by the introduced equations 3.1-17 (chapter 4). These parameters are taken to calculate the homogenised ply properties using equations 4.1-14.

6.3 Mechanical analysis module

The mechanical module should capture all relevant variations of material parameters in order to represent the material behaviour and the process itself. The mechanical part of the analysis can be divided into two steps, due to the nature of the process and the boundary conditions. In the first part, during curing, the composite part is fixed into a mould and therefore strong boundary conditions have to be applied. In the end of the process cycle, after cooling down to room temperature, the composite part is removed from the mould and the boundary conditions are changing. Before removal from the tool, large residual stresses appear which transform into residual deformations. The stress state of the composite part inside the mould and the removed part from the tool are quite different. Therefore, a strategy to take this into account is needed and will be derived in the following.

6.3.1 Step 1 - Curing process

The first step of the mechanical analysis module is focus on the curing process and the changes of the material in view of the development of the engineering constant, dependent on temperature and degree of cure. The observed composite material in this work is a unidirectional fabric material (G1157). This material can be idealised as an orthotropic material with 9 relevant material parameters. The second derivative of the deformation energy is the material tensor and can be written as follows

$$
C_{ij} = \begin{bmatrix}
C_{11} & C_{12} & C_{13} & 0 & 0 & 0 \\
C_{12} & C_{22} & C_{23} & 0 & 0 & 0 \\
C_{13} & C_{23} & C_{33} & 0 & 0 & 0 \\
0 & 0 & 0 & C_{44} & 0 & 0 \\
0 & 0 & 0 & 0 & C_{55} & 0 \\
0 & 0 & 0 & 0 & 0 & C_{66}
\end{bmatrix}. \tag{6.3}
$$

The resin module is changing during the process. Taking this into account, eq. 4.15-3.24 are used. The changes of the resin module are used to compute the homogenised properties on every time step. This is done for an orthotropic material and simplified to transverse isotropic behaviour by eq. 6.9. Using the engineering constants, the stiffness components of the material tensor are calculated by the following equations:

$$
C_{11} = \frac{1 - v_{23}v_{32}}{E_1 E_2 \lambda}, \qquad C_{12} = \frac{v_{12} + v_{31}v_{23}}{E_3 E_2 \lambda}, \qquad C_{13} = \frac{v_{13} + v_{21}v_{32}}{E_3 E_2 \lambda}, \tag{6.4}
$$

$$
C_{22} = \frac{1 - v_{13}v_{31}}{E_1 E_3 \lambda}, \qquad C_{23} = \frac{v_{32} + v_{12}v_{31}}{E_1 E_3 \lambda}, \tag{6.5}
$$

$$
C_{33} = \frac{1 - v_{12}v_{21}}{E_1 E_2 \lambda}, \tag{6.6}
$$

$$
C_{44} = G_{12}, \qquad C_{55} = G_{23}, \qquad C_{66} = G_{13}, \tag{6.7}
$$

$$
\lambda = \frac{1 - v_{12}v_{21} - v_{23}v_{32} - v_{13}v_{31} - 2v_{21}v_{32}v_{13}}{E_1 E_2 E_3}. \tag{6.8}
$$

The main factors for process-induced deformations are the chemical- and thermal-induced strains. Therefore, an incremental formulation of the strains are defined and added to the mechanical strain

$$
\Delta \varepsilon_{tot(p,T)} = \Delta \varepsilon_{el} + \Delta \varepsilon_{th(p,T)} + \Delta \varepsilon_{sh(p)}. \tag{6.9}
$$

The chemically-induced strains are based on the increment of the degree of cure. The total value of the vector quantity is the computed in a homogenisation approach equations 4.34-37

$$
\Delta \varepsilon_{sh(p)} = \gamma_i \Delta p. \tag{6.10}
$$

A similar approach is applied to the thermally-induced strains, also based on the presented homogenisation approach (equations 4.25-29)

$$\Delta\varepsilon_{th} = \alpha_l \Delta T .$$ (6.11)

The development descriptions, which take the relaxation behaviour for a variable temperature from RT to 200°C, a variable glass transition temperature and variable fibre volume content into account, have to be transformed in an incremental manner, which can be implemented via user subroutine of the implicit finite element program SAMCEF Mecano. The incremental form was developed by Zocher [28] and also used by Svanberg [25]. As a first step, the stress of the actual time step is a sum of the total stress and the stress increment

$$\sigma_{(t+\Delta t)} = \sigma_{(t)} + \Delta\sigma .$$ (6.12)

The stress increment is a sum of the derivative of the stiffness, multiplied with the total strain increment and a stress relaxation component. The strain increment consists of a sum of mechanical, thermal and shrinkage strain (eq. 6.9)

$$\Delta\sigma_{(t)} = \sigma^R + \Delta C \cdot \Delta\varepsilon_{tot(T,p)} .$$ (6.13)

The relaxation stress increment describes the decrease of the stress and consists of the following relation multiplied with the so-called recursive relation $S_{(t)}$, which is a state variable from the previous time step

$$\sigma^R = S_{(t)} \cdot \left(1 - e^{\frac{-\Delta\xi}{\rho}}\right).$$ (6.14)

The definition of the recursive element of the actual time step is defined as follows

$$S_{(t+\Delta t)} = e^{\frac{-\xi}{\rho}} \cdot S_{(t)} + \frac{\varrho \cdot C \cdot \Delta\varepsilon_{total}}{\Delta\xi}\left(1 - e^{\frac{-\Delta\xi}{\rho}}\right).$$ (6.15)

Accordingly, the full relaxation behaviour is based on three variables: ϱ, ρ and the shift factor a_t. The developed model is dependent on the description of the relaxation of pure resin. Furthermore, the following relation describes the resin behaviour, dependent on temperature:

$$\varrho = 0.0544 \cdot e^{0.0448 \cdot T} \quad [°C] ,$$ (6.16)

$$\rho = 35.638 \cdot e^{0.0134 \cdot T} \quad [°C] ,$$ (6.17)

$$a_T = -0.04983 \cdot T + 10.83 \ [°C] .$$ (6.18)

The parameters $p_{ij(p,T)}$ and $\varrho_{ij(p,T)}$ are temperature and degree of cure dependent parameters, respectively, to describe the relaxation behaviour. They are a vector quantity, therefore, six values have to be determined to describe the anisotropic relaxation behaviour. The complex behaviour can be simplified by the following assumption [25]:

- The viscoelastic behaviour is based on the matrix component
- The fibres have a linear elastic behaviour
- Homogenisation methods can be used to transfer the matrix relaxation behaviour to the composite behaviour.

Hence, the characterisation of the relaxation behaviour can be simplified to resin, which shows an isotropic behaviour. The isotropic material can be describe by the parameters $p_{(p,T)}$ and $\varrho_{(p,T)}$. The complete characterisation of the resin behaviour was presented in Chapter 3. Using the approach of Meder [58], the following homogenisation can be done:

$$r_1 = \frac{1}{1 + \dfrac{V_f}{(1 - V_f)\dfrac{E_m}{E_{f1}}}} , \tag{6.19}$$

$$r_2 = \frac{1}{1 + \dfrac{V_f}{(1 - V_f)\dfrac{E_{f2}}{E_m}}} . \tag{6.20}$$

This can be applied on the relaxation parameters $p_{(p,T)}$ and $\varrho_{(p,T)}$ for a transversal isotropic material:

$$\varrho_{11(p,T)_1} = \varrho_{(p,T)} \cdot r_1 , \qquad\qquad p_{11(p,T)} = p_{(p,T)} \cdot r_1 , \tag{6.21}$$

$$\varrho_{22(p,T)} = \varrho_{33(p,T)} = \varrho_{(p,T)} \cdot r_2 , \qquad p_{22(p,T)} = p_{33(p,T)} = p_{(p,T)} \cdot r_2 , \tag{6.22}$$

$$\varrho_{12(p,T)} = \varrho_{13(p,T)} = \varrho_{23(p,T)} = \varrho_{(p,T)} \cdot r_2, \tag{6.23}$$

$$p_{12(p,T)} = p_{13(p,T)} = p_{23(p,T)} = p_{(p,T)} \cdot r_2 . \tag{6.24}$$

6.3.2   Step 2 - Tool removal

The stress state for the situation inside the mould and after separation from the tool is quite different, because the boundary conditions are changing. Inside the mould, all displacements normal to the part surface are suppressed by the mould. This leads to high stress state which remains during demoulding and transforms to deformations and the final shape with a different state of residual stresses. Different studies of the mould boundary conditions can be found by Svanberg [25], Johnston [4], studies about the tool part interaction are done by Osooly [64] and Twigg et al. [65]. Most of these publications are related to the autoclave process and study experimentally and / or numerically, the influence of parameters like release agent, tool material autoclave pressure, layup on deformations.

There are two approaches to implement the influence of the tool. The first possibility is to model the tool in detail and applies different types of connection between part and mould like contact elements, shear layer or fixed connections. This increases the computational effort, especially, if the tool part interaction is modelled with contact elements.

This might not be applicable for large composite parts. The second approach is to define that the expansion of the mould does not have any influence on the composite part deformations, for example, if a closed mould is used with similar CTE. In this case the mould can be replaced by a translational fixation in normal direction of the surface to model the situation in the mould. During the demoulding process, this boundary condition, which is initially fixed, can be liberated at a certain time. As shown on the following Figure 60, once a given time is reached, the Degree Of Freedom (DOF) becomes free. The Lagrange multiplier, used initially to fix the degree of freedom, is removed.

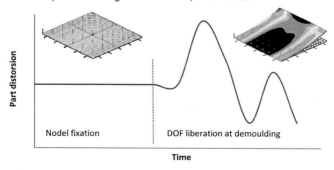

Figure 60 Schematic description of demoulding, degree of freedom liberation

A detailed study of the effect of different boundary conditions will be discussed in chapter 6.5

6.3.3   Post processing of the results

In general, a transient analysis method like the presented strategy is time consuming and has to be in balance with the value of benefit of additional information about the process. One advantage of the developed method is that there are not only deformation and stresses available as results. Using the method, also the changes of the engineering properties can be analysed over the process time and we end up with a total characterisation of the composite material on the ply level. Therefore, the following results are available:

Table 6.1

Pos	Parameter name	Explanation	Symbol	Unit
1	p	Degree of cure	p	-
2	dcdt	Reaction velocity	dp/dt	1/min
3	Tg	Glass transition temperature	T_g	°K
4	Eh1	Module in fibre direction	E_1	Pa
5	Eh2	Module in transverse direction	E_2	Pa
6	Eh3	Module in out of plane direction	E_3	Pa
7	Em	Module resin	E_m	Pa
8	Gh12	Shear module 12	G_{12}	Pa
9	Gh13	Shear module 13	G_{13}	Pa
10	Gh23	Shear module 23	G_{13}	Pa
11	poi13	Poisson ratio 13	v_{13}	-
12	poi12	Poisson ratio 12	v_{12}	-
13	poi23	Poisson ratio 23	v_{23}	-
14	alh1	CTE fibre direction	α_1	1/K
15	alh2	CTE transverse direction	α_2	1/K
16	alh3	CTE out of plane direction	α_3	1/K
17	beta1	CSC fibre direction	β_1	1/%p
18	beta2	CSC transverse direction	β_2	1/%p
19	beta3	CSC out of plane direction	β_2	1/%p

Additionally of course, deformations, strains, and stresses are available as results. To analyse the residual stress state existing yield criteria like Von Mises, Tresca, Drucker/Prager are not applicable for materials with anisotropic strength values. Comparing all these stress components is very time consuming and it is not applicable for large structures, because there is no information about interaction between the different stress components. Especially in the case of a curved structure like an L shape angle, the out of plane component is important because it can lead to failure by delamination [66]. Therefore, the 3D version of the Puck criterion was added to compute the fibre effort ε_{FF} and the inter-fibre effort ε_{IFF} during the process to condensate the information of the stress tensor to one scalar value. During the process, all strength values develop to their cured value. The application of the Puck criterion on the case is performed with the strength values of the cured material [29]. It is a quantitative declaration of the interaction of the stress components and not a failure of the material during the process. The implementation of the Puck criterion was done by the author in the previous work and is published [67]. Following equations are used to compute the fibre effort and the inter fibre effort.

Material effort condition	Area of validity
$\varepsilon_{IFF} = \sqrt{\left[\left(\dfrac{1}{R_\perp^+} - \dfrac{p_{\parallel\psi}^+}{R_{\parallel\perp}}\right)^2 \sigma_n^2 + \left(\dfrac{\tau_{nt}}{R_{\perp\perp}^A}\right)^2 + \left(\dfrac{\tau_{n1}}{R_{\parallel\perp}}\right)^2\right]} + \dfrac{p_{\parallel\psi}^+}{R_{\perp\psi}^A}\sigma_n$	$\sigma_{22} > 0$
$\varepsilon_{IFF} = \sqrt{\left[\left(\dfrac{\tau_{nt}}{R_{\perp\perp}^A}\right)^2 + \left(\dfrac{\tau_{n1}}{R_{\parallel\perp}}\right)^2\right]} + \dfrac{p_{\parallel\psi}^-}{R_{\perp\psi}^A}\sigma_n$	$\sigma_{22} < 0$

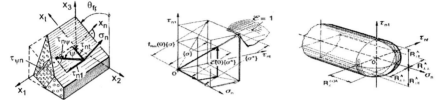

Figure 61 Stress on the fracture plane[66]   Vector of material effort[66]   Material effort body[66]

6.4 Validation test case 1

In the first validation case, the developed material model is compared against experimental values of engineering properties and strength values. The model provides results for the degree of cure, the glass transition temperature, resulting engineering properties and process-induced stress. To validate these values, two plates are cured at different temperatures, characterised and compared against the simulation results.

6.4.1 Experimental investigation

In this experiment, the objective is the characterisation of the influence of process-induced stresses on strength and stiffness values. Using a phenomenological approach, test specimens are manufactured with different levels of process-induced stresses and tested in tensile tests. Two different test plates are manufactured using RTM6 resin and 8 layers of CFRP unidirectional fabric (G1157). The first plate was cured using the standard conditions. A curing temperature of 180°C for 120 min was used. The second plate was cured at 120°C for 300min. Therefore, the first plate has a higher amount of residual stresses. For each plate, 10 specimens are tested in fibre direction and transverse to the fibre direction. All tests were performed using the standard norm DIN EN 527-2. Table 6.2 shows the results of the tests [S6]:

Table 6.2 Results in fibre direction	$R_{\parallel}$ [N/mm^2]	Standard Dev. [N/mm^2]	E_1 [N/mm^2]	Standard Dev. [N mm^2]
Curing temperature 180°C	1746.4	58.0	124504.0	3082.3
Curing temperature 120°C	1904.0	32.6	123355.9	4488.6

Results transverse to the fibre direction	$R_{\perp}$ [N/mm^2]	Standard Dev. [N/mm^2]	E_2 [N/mm^2]	Standard Dev. [N/mm^2]
Curing temperature 180°C	33.1	4.37	8367.8	91.6
Curing temperature 120°C	39.9	2.10	8773.1	155.6

Results in 45° direction	$R_{\parallel\perp}$ [N/mm^2]	Standard Dev. [N/mm^2]	G_{12} [N/mm^2]	Standard Dev. [N/mm^2]
Curing temperature 180°C	35.3	3.02	4702.5	64.8
Curing temperature 120°C	35.6	2.98	4399.7	72.5

The interpretation of the experimental results leads to the following conclusion. In the case of the transverse applied load, high process-induced stresses have a significant influence and degrade the transverse strength. In the case of the applied load in fibre direction, the strength values show a different behaviour, but it has to be mentioned that the standard deviation is high and a reliable conclusion cannot be found. In the case of transverse direction, the stiffness increases if the internal stresses decrease, and in the case of fibre direction, the stiffness increases also if the internal stresses decrease.

This experimental investigation can be used to evaluate quantitatively, the level of process-induced stress. Using the difference in the strength values of the different curing temperatures, a stress value can be computed. This leads to the following stress values:

Table 6.3 Results – process-induced stress	120°C – RT [N/mm²]	180°C – 120°C [N/mm²]	180°C – RT [N/mm²]
Stress in fibre direction	262.6	157.6	420.2
Stress in transverse direction	11.3	6.8	18.3
Shear stress	0.5	0.3	0.8

The computed stress values are a rough estimation of the stress level because no nonlinear effects are taken into account. The stress in fibre direction is about 262N/mm² (120°C curing) and 420N/mm² (180°C curing). The stress in transverse direction is 11.3N/mm² (120°C curing) and 18.3N/mm² (180°C curing). The influence on the shear stress is not visible. This is explained by chemical and thermal shrinkages, which are always applied in normal direction. The detailed study about the formation of stress on the micro scale has shown that several nonlinear effects such as relaxation, plasticity, degradation influence the stress state and resulting properties on the ply level (meso scale), significantly. Therefore, the stress values evaluated form the experiment are just an approximation.

6.4.2   Numerical investigation

The objective of the present study is a comparison between a coupled process mechanical simulation and the experimental tensile test. In this study, different validations are presented by the following parameters: final degree of cure, homogenized engineering constants and the tensile strengths. In this validation study, first a process simulation of the curing process will be done to get the internal stress state and second an analysis of the tensile test. The FE-model was built using the average geometries of the specimens.

Table 6.4 Dimensions of the investigated specimen

Width (Average)	w = 25.03 mm
Thickness (Average)	t = 2.183 mm
Length	l = 130 mm

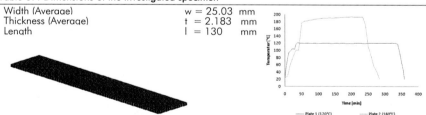

Figure 62 FE discretisation of specimen and temperature load

Starting from this geometry a parametric FE - model was created, consisting of 3380 hexahedron composite volume elements using 4 elements over the thickness (Figure 62). In the FE - model, different mechanical boundary conditions are applied. During the curing process simulation step, the model is constrained on all faces in normal direction.

These boundary conditions have to change in the tensile test part, where a displacement function is applied to the face of the right side and a complete clamp is applied to the opposite side. The right boundary conditions for the curing conditions are quite difficult to apply. The plate is connected to an aluminium plate during curing and cooling. After demoulding the plate deforms to the actual shape and some stress releases. This situation is modelled with the previous described strategy with liberation of degrees of freedom during the analysis.

For each test case (120°C and 180°C) the fracture displacement of 0.55mm and 0.43mm was applied. A temperature load was applied using the temperature profiles of the process, which are taken from a thermo couple measurement placed in the middle of the thickness of the plates. By analysing the temperature distribution, it is visible that exothermic heat flux does not have an influence (constant temperature) and a thermodynamic analysis can be neglected. Therefore, the temperature profiles of Figure 62 are applied on the model.

Similar to the experimental data, a stacking of 8 CFRP layers with a thickness of 0.27mm per layer and a total thickness of 2.18mm was modelled. Using the developed material model all material values based on the experimental investigation of chapter 3 and 4.

The first comparison was done on the total degree of cure. The experimental degree of cure was determined using dynamic DSC (Differential Scanning Calorimetric) runs and measuring the residual exothermic reaction enthalpy. In proportion to the total reaction enthalpy the residual degree of cure was computed [39].

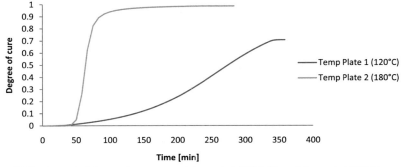

Figure 63 Development of degree of cure, green-plate cured at 180°C, blue- plate cured at 120 °C

Table 6.5 Comparison - degree of cure

Process simulation	Experiment	Difference
$p_{120} = 0.71$	$p_{120} = 0.73$	2.7%
$p_{180} = 0.98$	$p_{180} = 0.96$	2.4%

The residual degree of cure evaluated in the process simulation fits with the experimental values. The numerical results are match in between the standard deviation of the experimental test series (around 3% [11]).

Results - Comparison of the glass transition temperature
The next validation is performed on the glass transition temperature. The experimental values are similar to the residual degree of cure determined by a DSC experiment.

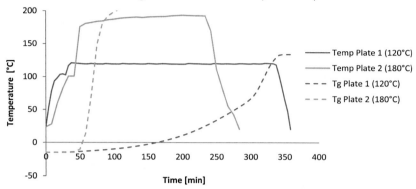

Figure 64 Development of glass transition temperature, green-plate cured at 180°C, blue- plate cured at 120 °C

Table 6.6 Comparison Glass transition temperature

Process simulation	Experiment	Difference
$T_{g\,120} = 132°C$	$T_{g\,Av\,120} = 137.7°C$	3.8%
$T_{g\,180} = 214°C$	$T_{g\,Av\,180} = 204.6°C$	-4.9%

In comparison to the experimentally evaluated glass transition temperature, the numerical results have a deviation of 3.8% (120°C) and -4.9% (180°C). This might look very large, but one has to keep in mind that measuring the glass transition temperature has a large standard deviation also dependent on the measuring method (DSC vs. DMA).

Results - Comparison of homogenised engineering constants
The comparisons of the homogenised engineering constants are done for the Young's modulus in fibre direction and transverse to the fibre direction. First, Figure 65 presents the development of modulus in fibre direction and in transverse direction for the curing condition of 120°C over the degree of cure. The modulus in fibre direction is divided by a factor of 10 to implement the result into the diagram.

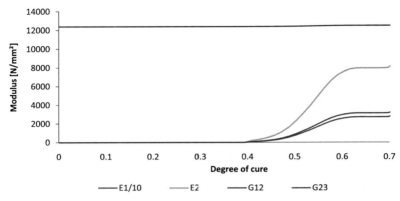

Figure 65 Development of the eng. constants on the ply level, curing condition at 120°C

Figure 65 shows the development of the composite properties for the curing at 120°C over the degree of cure. It is visible that the derived function eq. 3.20 has nonlinear dependence on the degree of cure. The development starts at the gel point around a degree of cure of 0.4 and goes up until the vitrification point is reached which is due to the low curing temperature at a low degree of cure around 0.61.

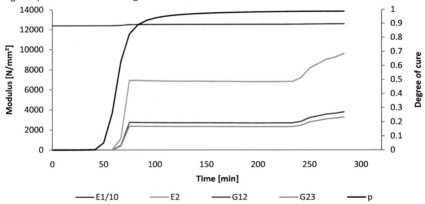

Figure 66 Development of the eng. constants on the ply level during process time, curing condition at 180°C

The diagram Figure 66 shows the development of the modulus over the process time for the curing condition at 180°C. The changes of the modulus due to the curing (from 50-90min) and the temperature (during cooling 220-280min) are visible. The influence of the modulus in fibre direction is less but the changes of all transverse properties are high. The sensitivity of the changes dependent on curing and temperature are both included and it is visible that neglecting one of them leads to wrong stress values. Figure 67 shows similar properties for the curing condition of 120°C.

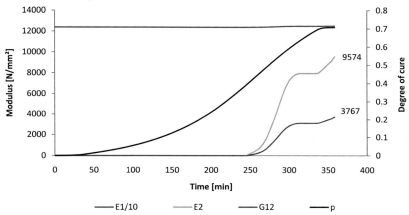

Figure 67 Development of eng. constants on the ply level during process time, curing condition at 120°C

Table 6.7 Comparison modulus in fibre direction

Process simulation	Experiment	Difference
$E_{1\,120} = 125255\ N/mm^2$	$E_{1\,120} = 123355\ N/mm^2$	1.5%
$E_{1\,180} = 125466\ N/mm^2$	$E_{1\,180} = 124504\ N/mm^2$	0.7%

Table 6.8 Comparison modulus transverse to the fibre direction

Process simulation	Experiment	Difference
$E_{2\,120} = 9574\ N/mm^2$	$E_{2\,120} = 8773\ N/mm^2$	9.1%
$E_{2\,180} = 9599\ N/mm^2$	$E_{2\,180} = 8367\ N/mm^2$	12.8%

Comparing the experimental and simulated values for stiffness in fibre direction and transverse to the fibre, a good agreement can be found. Most of the differences are smaller than the standard deviation of the experiment. Only the transverse modulus of the specimens cured by 180°C show a difference of more than 12.8%. In the experiment, an interesting effect is visible in the Young's modulus of low temperature cured specimens, the modulus is higher than the high temperature cured specimens. The simulation results show a different trend and the modulus is lower for low temperature cured specimens. Table 6.9 shows a comparison between process simulation results of the higher temperature cured specimens (180°) and the analytical approaches of Hashin / Hill.

Table 6.9 Comparison Num / Analytical values with initial fibre volume content

Process simulation	Hashin / Hill	Difference	
$E_1 = 125466 \, N/mm^2$	$E_1 = 133379 \, N/mm^2$	6.3%	
$E_2 = E_3 = 9599 \, N/mm^2$	$E_2 = E_3 = 11015 \, N/mm^2$	14%	
$G_{12} = G_{13} = 3788 \, N/mm^2$	$G_{12} = G_{13} = 4359 \, N/mm^2$	32%	
$G_{23} = 3285 \, N/mm^2$	$G_{23} = 3775 \, N/mm^2$		14 %
$v_{12} = 0.27$	$v_{12} = 0.26$	3.8%	
$v_{23} = 0.46$	$v_{23} = 0.45$	2.1%	
$V_f = 0.63$	$V_f = 0.63$		

It has to be mentioned that the fibre volume content changes due to chemical and thermal shrinkage during the process. The input fibre volume content was given by 0.63. After the curing process, this value was 0.61. This changing of the fibre volume content explains the difference between analytical and numerical values. The following table 6.10 shows the comparison using a corrected analytical fibre volume content value.

Table 6.10 Comparison Num / Analytical values with corrected fibre volume content

Process simulation	Hashin/Hill	Difference
$E_1 = 125466 \, N/mm^2$	$E_1 = 129237 \, N/mm^2$	3.0 %
$E_2 = E_3 = 9599 \, N/mm^2$	$E_2 = E_3 = 10415 \, N/mm^2$	8.8 %
$G_{12} = G_{13} = 3788 \, N/mm^2$	$G_{12} = G_{13} = 4106 \, N/mm^2$	8.3 %
$G_{23} = 3285 \, N/mm^2$	$G_{23} = 3567 \, N/mm^2$	8.5 %
$v_{12} = 0.27$	$v_{12} = 0.26$	3.7 %
$v_{23} = 0.46$	$v_{23} = 0.46$	0.0 %
$V_f = 0.63$	$V_f = 0.61$	

The comparison highlights how sensitively the fibre volume content influences the material properties. The resulting fibre volume content is dependent for open mould process on vacuum pressure and shrinkage. The experimentally determined value of the fibre volume content was 0.63 with a standard deviation of 5.3%. Especially, the high standard deviation of 5% is critical. Nevertheless, the numerical results show a good agreement with the experimental values.

Results - Comparison of homogenised coefficient of thermal expansion
The CTE was measured using a TMA method with a compression setup in the range of 0°C to 200°C in chapter 4. As shown previously, the CTE of the resin is not a constant and is dependent on the temperature. This will influence also the CTE of the composite. In the following curve, the CTE in fibre direction and transverse direction is shown over the process time for both curing conditions. There is an interesting effect visible. The temperature dependency of the CTE especially in transverse direction forces the thermal shrinkage

because the maximal CTE2 for the curing condition $18\,°C$ is $30.0 \cdot 10^{-6}/°C$ and for the curing condition of $120°C$ $25.4 \cdot 10^{-6}/°C$. This is an increase of around 16%.

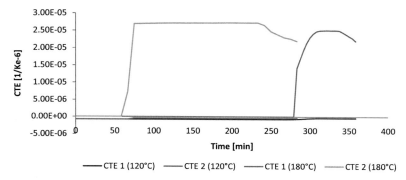

Figure 68 Changing of the CTE 1 and 2 during process for both curing conditions

Table 6.11 Comparison CTE

Process simulation	Experiment	Difference
$\alpha_{1\,120} = -0.4 \cdot 10^{-6}/°C$	$\alpha_1 = -0.3 \cdot 10^{-6}/°C$	25%
$\alpha_{1\,180} = -0.4 \cdot 10^{-6}/°C$		
$\alpha_{2\,120} = 21.9 \cdot 10^{-6}/°C$	$\alpha_2 = 23.8 \cdot 10^{-6}/°C$	8.6%
$\alpha_{2\,180} = 21.9 \cdot 10^{-6}/°C$		

Standard Deviation on total average (table 4.4)

The results of a comparison of the experimental and numerical evaluated values achieved for room temperature are presented in Table 6.11. The accordance is quite low and it is visible that, due to the effects of nonlinearity based on temperature and fibre volume content on one side, and measurement with a high standard deviation, an uncomfortable situation is present which has a large impact of the accuracy.

Results - Comparison of stresses
The comparison of the stress values is done as follows. First, a calculation without relaxation (full lines) and relaxation (dashed lines) is presented for both curing conditions and the tensile experiment. The stress values which are shown are the maximal values. Without relaxation the simulation results are overestimated. The difference with the experimental values can be used to evaluate the effect of relaxation. Figure 69 shows the development of the stress. The green curve displays the result of the plate cured at 180°C, the blue curve shows the plate cured at 120°C.

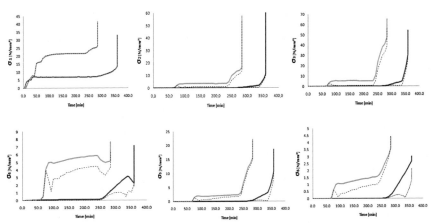

Figure 69 Development of process-induced stress and mechanical-induced stress, stress tensor components, green-plate cured at 180°C, blue-plate cured at 120°C

The development of stress by component over the process time shows a complex behaviour. First of all, it is visible that the stress in fibre direction does not relax but all other stress components do. The total stress can be divided into three types: chemical-dependent stress, thermal-dependent stress and mechanical- dependent stress due to externally applied deformations. Figure 70 show the development of stress for the stress component in transverse direction in detail.

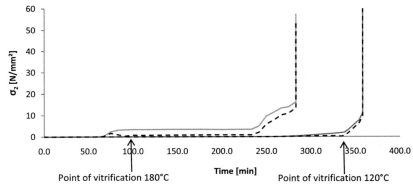

Figure 70 Development of process-induced stress and mechanical-induced stress of stress component 2 (transverse direction), green-plate cured at 180°C, blue-plate cured at 120 °C, for IE and VE solution

It is visible that the chemical induced shrinkage occurs after the gel point and increases strongly during the rubber phase. The incremental elastic solution and the viscoelastic solution show a large deviation in this phase. It is visible that the chemical-induced stress is relaxing fast in the rubbery phase and, after reaching the point of vitrification the relaxation slows down. For the curing situation of 120°C, the point of vitrification is reached late, just before cooling. Therefore, the chemical-induced stress is nearly relaxed. During the cooling, the incremental elastic solution and the viscoelastic solution are nearly similar. The total amount of process-induced stress for the curing condition of 120°C is 11.2N/mm² for the IE solution and 9.7N/mm² for the VE solution. The total amount of stress for the curing condition 180°C is about 16.2N/mm² for the IE solution and 13.2N/mm² for the VE solution. Compared to the experimentally evaluated values (table 6.3) these stress values are all lower, but it has been taken into account the experimental values are a linear approximation.

Comparing all these stress components is very time consuming and it is not applicable for large structures. There is no information about interaction between the different stress components. Analysing the stress components of Figure 69 has shown that the out of plane stress component have the same magnitude as the transverse component and influence the failure behaviour significantly. Therefore, the Puck criterion was added to compute fibre effort and the inter-fibre effort during the process to reduce the information of the complex stress tensor to one scalar value. During the process, all strength values develop to their cured value. The application of the Puck criterion on the case is performed with the strength values of the cured material [60]. It is a quantitative declaration of the interaction of the stress components and not a failure of the material during the process. Figure 71 shows the inter-fibre effort value over time and degree of cure.

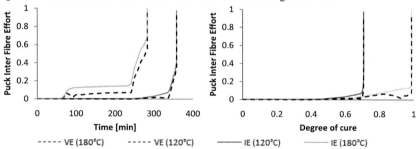

Figure 71 Development of process-induced and mechanical-induced inter fibre effort, green-plate cured at 180°C, blue-plate cured at 120 °C, for IE and VE solution

Table 6.12

Incremental elastic (IE)	chemical induced	thermal induced	total process induced	mechanical induced	Total
$\varepsilon_{IFF\ 120°C}$	0.07	0.34	0.40	0.58	0.97
$\varepsilon_{IFF\ 180°C}$	0.14	0.53	0.68	0.39	1.07
Viscoelastic (VE)					
$\varepsilon_{IFF\ 120°C}$	0.01	0.33	0.34	0.56	0.92
$\varepsilon_{IFF\ 180°C}$	0.07	0.48	0.55	0.39	0.95

The previous table 6.12 determines the differences between the two curing conditions and the results of the incremental elastic solution and the viscoelastic solution. The material inter-fibre efforts of the process-induced stresses show large differences between the two curing conditions. In the case of low curing temperature the chemical-induced effort is reduced from 0.07 to 0.01, the thermal-induced effort is decreasing form 0.48 to 0.33 and the total sum of the effort decreases from 0.55 to 0.34 for the viscoelastic solution.

Conclusion

In the present study, process-induced residual stresses of composite materials have been analysed using a process simulation of a unidirectional laminate in comparison to experimental tests with two differently cured plates. The following results were found:

- Using a tensile test, an influence of process-induced stresses can be found, related to the variation in curing temperature.
- The developed simulation method can be used to analyse the manufacturing process and provides the following results with reliable accuracy: total degree of cure (<3%), resulting glass transition temperature (<5%), resulting engineering constants (<9%), resulting coefficient of thermal expansion (<25%), process-induced stresses (no reliable experimental data available).

6.5 Validation test case 2

The second validation test case is focused on the correct representation of warpage. The previous validation was focused on representing the correct material behaviour with a focus on the transient changes of the engineering properties. In this validation, another aspect is important. Process-induced deformations are amplified in case of an unsymmetrical laminate stacking or effects which disturbed the symmetrical conditions in a stacking like preform errors, draping errors etc. This leads to the so-called bimetal effect if a temperature load is applied. The important stiffness components in this case are the transverse shear stiffness components. Therefore, the finite elements which are used in a numerical analysis have to be proved to represent this transverse shear stiffness and shear deformations in the right way. As presented in the simulations strategy, shell elements should be used to ensure the application to large structures. Normal shell elements based on the theories of Love [68] and Kirchhoff [69] and do not take into account transverse shear deformation. Consequently, shell elements with a Reissner Mindlin [70] approach have to be used. Consequently, the second validation test case will be dedicated to test different element types.

6.5.1   Experimental investigation

In this experiment, the objective is the characterisation of process-induced deformations. Using a phenomenological approach, test specimens are manufactured with an unsymmetrical laminate stacking of [0₃/90₃]. The plates are manufactured using RTM6 resin and 6 layers of CFRP unidirectional fabric (G1157). To ensure right conditions, two plates have been cured at the same time and this was repeated 5 times. Furthermore, a total of 10 plates was the database to measure process-induced deformations. The measurement of the deformations was performed using a simple method with a calliper.

Due to the nature of the manufacturing process conditions, different disturbance are influencing the process and, consequently, the quality of the composite part. Figure 72 shows the temperature profiles measured in the composite part. The parts have been cured in a standard oven, hence, variations of curing temperature, heating rate, cooling rate, curing time are always present and listed in table 8.2.

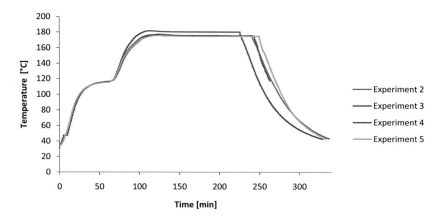

Figure 72 Temperature variations for 4 curing cycles

This highlights that a deterministic simulation approach might not be sufficient to answer all questions about part quality. Therefore, in Chapter 8 the influence of parameter variability will be discussed in detail. For the given case, average values were used and compared to the simulation results. Figure 73 presents the deformation of the plate in the out of plane direction (Z direction) over the dimension of the width, for the left and the right edge of the plate. In this figure, the values for all 10 plates are shown as points and fitted with a polynomial trend line of second order. The average deflection is 36.29mm (standard deviation 1.9mm) [S4].

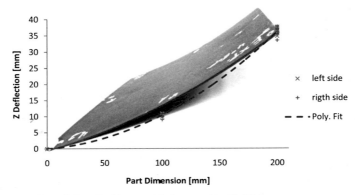

Figure 73 Measured out of plane Z - defection of 8 plates, stacking [0₃/90₃]

6.5.2   Numerical Investigation

The objective of the present study is a comparison between numerical analysis with different element types and the experimental data of the deformed shape of the plate. The FE-model was built using the average geometries of the specimens (Table 6.14). Starting from this geometry, three parametric FE - models were built up, consisting of 40 hexahedron composite volume elements using 1 element over the thickness (Figure 74), 40 hexahedron composite volume elements using 1 element over the thickness and 40 linear quadrangle composite shell elements. The shell elements are of high order, including a Reissner Mindlin kinematic.

Table 6.14 – Dimensions of the investigated specimen

Width (Average)	a = 208.68 mm
Length (Average)	b = 208.38 mm
Thickness (Average)	t = 1.66 mm

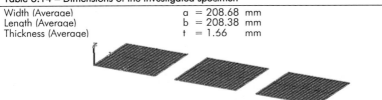

Figure 74 FE discretisation

A temperature load is applied using an average function of the measured temperature profiles (Figure 72). The material properties data were taken from the previous chapter 3. The elements are tested in a linear static analysis, a geometrical nonlinear analysis and a nonlinear analysis with the developed material model. The plate is clamped during the curing process in all normal face directions up to the point there room temperature was reached. In the second step of the analysis, the boundary conditions are changed to simulate the removal process and to get the right stress state after tool removal. The following

Figure 75 shows out of plane displacement as scalar value for a viscoelastic solution after tool removal step.

Figure 75 FE result out of plane defection, 37.53mm

Table 6.15 summarise the out of plane deformation for a linear and nonlinear analysis and the corresponding deviation to the experimental result. In case of a linear analysis it can be conclude, that the occurring deformation do not fit to the experimental results independent from the element type. The second case taking into account geometrical nonlinearities but linear material behaviour and shows that the occurring deformations for all types of volume elements have a negative deviation (-22%) and the shell element have a positive deviation (12%). The deviations are also not acceptable for this kind of analysis. The negative deviation is the expected behaviour because this type of analysis takes only the thermal shrinkage into account. The chemical shrinkage is not considered. The third analysis includes the chemical induced shrinkage and all nonlinear material effects using the developed material model. The volume elements of type composite volume shell and composite volume show a good approximation to the experimental value by a deviation of 3.3 and 3.6%.

Table 6.15 – Out of plane displacement, measured on the corner node

Element type	linear	Deviation	Geometric non linear	Deviation	Geo. + material model	Deviation
Composite Volume	45.31m	24%	28.21 mm	-22%	37.53mm	3.3%
Composite Volume	51.95m	43%	28.22 mm	-22%	37.65mm	3.6%
Reissner Mindlin	51.77m	43%	40.62 mm	12%	73.8	103%

The shell element shows a high deviation of around 103%. The reason for this high deformation in both analyses can be found in the theory of the shell element. Shell elements are generally sensitive in view of high transverse shear stress. In case of a laminate with a high degree of unsymmetrie [0₃/90₃] the assumption of the shell element theory is not ful-filled. One of these restrictions is that the shell cross section must be perpendicular to the neutral surface of the shell element. This assumption is not fulfilled if transverse shear ef-fects like the deformations forced by different CTE over the thickness are dominant. An-other approximation of shell theory is that the transverse shear stresses are assumed being parabolic distributed over the thickness. For an isotropic material this assumption is nearly valid and to obtain more accuracy a shear correction factor is introduced. In case of lay-ered materials with a symmetric layup the parabolic shear distributions is different which is illustrated in Figure 76. The shear correction factor is a fixed value implemented in the shell elements. Therefore, a user defined shear correction dependent on the laminate se-quence is not possible. Consequently, the shell elements are only applicable for symmetric laminates [2, 85].

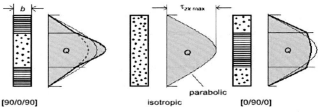

Figure 76 Transverse shear stress distribution for a isotropic material, a [0/90/0] and a [90/0/90] laminate [2]

Accordingly, the applicability of shell elements is only valid for composite laminates with a low degree of unbalance or asymmetry. A criterion allowing assessing the potential influ-ence of the transverse shear deformation is given by the ratio [71]

$$\rho = \frac{E}{G}\left(\frac{t}{L}\right)^2 \ll 1. \tag{6.24}$$

For this formulation E is the Young's modulus of the laminate E_z, G are the transverse stiff-ness G_{yz}, t is the laminate thickness and L is a characteristic length over which the stress changes significantly. For the given laminate, the relation has a value of $\rho = 2.42$. Shear effects are dominant if ρ is bigger than 1. Therefore, the CLT assumption does not con-form. There are two possibilities in that case. The first possibility is to use solid elements, The second one is to use high order shell elements which feature interlayer rotational de-gree of freedom and a user defined shear correction factor. In the used software, the Mindlin elements are of type second order. There are developments of Bischoff [72], Ramm [73], Reddy [74] for high order shell elements, but they are not implemented into commercial FE codes and are more-or-less ongoing research work.

Additional to the deformation, the stress and the inter fibre effort can be analysed and displayed over the process time. Again, to reduce the information of the complex stress state, the inter fibre effort value of the Puck criterion is used to simplify the stress state to a scalar value. In the diagrams the maximum and minimum values of the viscoelastic solution and the incremental elastic solution are presented. Figure 77 to Figure 80 present the distribution of stress transverse to the fibre and the inter fibre effort value before demoulding and after demoulding for the viscoelastic solution.

Bevor demoulding

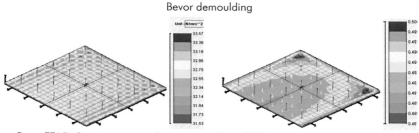

Figure 77 VE - Stress in transverse direction   **Figure 78** VE - Puck IFF value - min 0.48 max 0.57

After demoulding

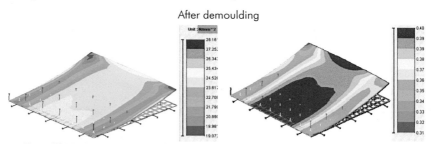

Figure 79 VE - Stress in transverse direction   **Figure 80** VE - Puck IFF value - min 0.30 max 0.43

Figure 77 and Figure 79 show the transverse stress before demoulding and after demoulding. It is visible that before demoulding the stress is nearly uniform distributed of the plate with a variation from 31.5 MPa to 33.5MPa. After demoulding the high stress transforms to process-induced deformations and changes to a no uniform distribution with values between 19 MPa to 28MPa. In Figure 78 and Figure 80, a verification of the boundary condition is made. Figure 81 presents the inter fibre effort value development during the process with a free standing condition. In this case, the plate is clamped to deform free during the curing process without any constraints which are normally imposed by the mould. In the second case, all nodes are clamped in normal direction to the face and released after the process.

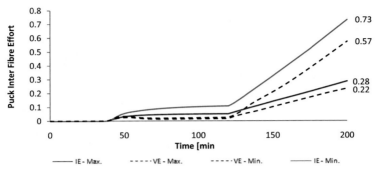

Figure 81 FE result inter fibre effort (min and max value) for the IE and VE solution, boundary conditions free standing curing

The diagrams in Figure 81 presents the development of stress and effort or the incremental elastic (IE) and the viscoelastic solution (VE) for a free standing curing condition. The relaxation effect has an important influence on the chemical-induced stresses and the chemical-induced shrinkage part is nearly zero. This corresponds to statements which can be found in the literature by White et al. [19], Wenzel [53], Wijskamp [24] and Zobeiry [16]. The range of effort for the incremental solution is from 0.28 to 0.73, the range of values for the viscoelastic solution from 0.22 to 0.57. This highlights that using a transient analysis schema the effect of viscoelasticity cannot be neglected and pure elastic solution overestimates the residual stress state by 12%. Figure 82 shows the same analysis with a fully constrained boundary condition until the end of the process and a DOF liberation (demoulding). This situation is like a total bond to a rigid mould without thermal contraction.

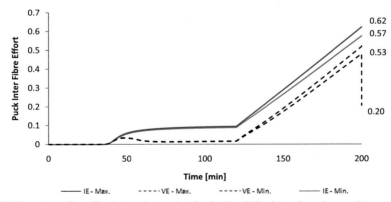

Figure 82 FE result inter fibre effort (min and max value) for the IE and VE solution, boundary conditions with tool releasing

Figure 82 indicates the solution for the incremental elastic (IE) and viscoelastic (VE) cases. In the case of using incremental elastic, the removing of the boundary condition leaded to convergence problems and no solution was found. In the case of the fixed boundary condition the minimum and maximum value during the process are nearly equal by 0.47 to 0.51. The stress state changes dramatically during demoulding to 0.20 to 0.53 and goes nearly to the range of value of the free standing curing condition (0.22 to 0.57). The maximum value is decreasing in case of free standing cure or demoulding from 0.57 to 0.53. This leads to following conclusion about the influence of boundary conditions. They have a large influence on the stress values during the process and also on the residual stress values after demoulding. Similar results were found by Svanberg [25].

6.6 Conclusion

In the present study, process-induced deformations of composite materials have been analysed using a process simulation of an unbalanced laminate in comparison to experimental tests. The following results were found:

- The performed validation case with the unbalanced stacking was suitable for showing the influence of the thermal and chemical induced shrinkages on process-induced deformations.
- The preferred solution for volume elements are composite volume shell elements.
- The applicability of shell elements is restricted if transverse shear effects are not to dominate and deflection is not large. This limits the application to symmetrical laminates or laminates with slightly disturbances of the symmetric conditions.
- Boundary conditions have an influence on the stress values during the process and on the residual stress values after demoulding if viscoelastic approaches are used.

7   Application - composite multi spar flap

In chapter 7 the validated method is applied to a test structure, namely to analyse the RTM manufacturing process of an integral Composite Multispar high-lift Flap (CMF). For highly integral monolithic structures, a major challenge is the development of a robust manufacturing process to produce high quality structures. An integral structure must be fitted to the tolerance requirements, because there is no possibility to change the final geometry in the assembling step. Process-induced deformations can be a risk factor for these types of structures in view of required tolerances, manufacturing costs and process time. The test structure has the length of 7.5m and consists of an unsymmetrical aerodynamic profile with five integrated spars. The application of the analysis method in the industrial environment on this structure presents the possibilities using parameter studies on process parameters to reduce process time, to evaluate risk parameters and their influence, and to improve part quality.

Figure 83 Cross section of the composite part

Figure 85 Heat press and mould

Figure 84 Composite part

The composite part is cured in a one-shot RTM process. Dry textile preforms are placed into an aluminium mould, then heated to the injection temperature of about 100°C, injected by liquid resin, heated to the curing temperature of 180°C, cured for 2h and cooled to room temperature. The part is removed from the tool in hot conditions therefore tool part interaction can be neglected in this case.

Due to the nature of a composite manufacturing process there are always two types of configurations which are important if a process like this has to be analysed. There will be a so-called "As-Planned" configuration and an "As-Built" configuration. During the development process of the manufacturing process a lot of uncertainties exist. A simulation method can support the development of the process by evaluation of the sensitivity of different process parameters to determine the specific tolerances for the parameters. On the other hand, existing processes can be analysed to identify the critical parameters and to develop strategies for the compensation of process-induced deformations.

In the following a FE - model is built on the "As-Built" configuration and different parameters like thickness deviation, orientation errors etc. have been added. The presented model is based on the parameters of the seven prototype of the CMF component. The aim of this study is to validate the numerical model, to perform a sensitivity analysis to identify the critical parameters and to give suggestion for the compensation of warpage.

7.1 Model description

Starting from an imported CAD model, the FE - model is built with a volume mesh using 985640 tetrahedral volume elements for the thermo analysis part of the sequential coupled analysis. The complete mould and the composite are meshed with volume elements.

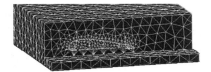

Figure 86 FE discretisation mould for the thermal analysis

As boundary conditions, a temperature profile (Fig. 87) was applied to the upper and lower sides of the mould. On the sides of the mould a natural convection is applied with a convection coefficient of $12W/m^2$ and a fluid temperature of $20°C$.

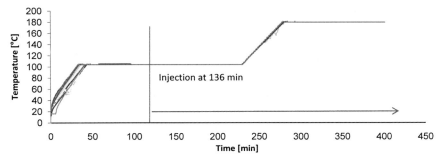

Figure 87 Temperature boundary conditions and infusion point

The mechanical FE - model of the flap is simplified to a shell structure. The laminate stacking is a quasi-isotropic laminate which consists of different types of fabric. Every fabric ply (0/90) is modelled using four unidirectional plies (0/90/90/0) similar to the suggested approach of Johnston [5]. The stacking is a balanced symmetrical laminate. As aforementioned, the "As-Built" configuration is analysed. Therefore, variations of the thickness, draping and layup errors are implemented into the FE - model. The model of the multispar flap was simplified to a shell structure presented in Figure 88

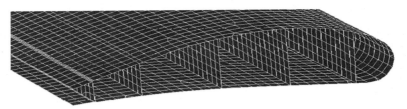

Figure 88 FE discretisation composite part for the mechanical analysis

The mesh of the flap consists of 1984 (8%) triangle and 23028 (92%) quadrangle multilayer linear shell elements with a Reissner Mindlin approach. The model is clamped in the middle of the structure.

7.2 Results

In the following, the results of the thermal and mechanical analyses steps are presented. From the thermal analysis module the temperatures inside the composite part and mould are displayed in Figure 89. The points ITL n are arbitrarily chosen. The dashed lines are representing nodal temperatures in the spars, the continuous lines nodal temperatures in the skin.

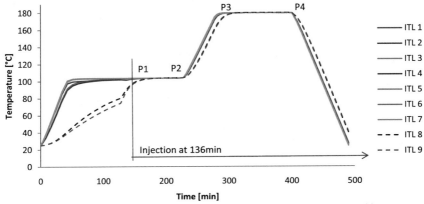

Figure 89 Temperature boundary conditions and temperature inside the composite part/mould

In the beginning of the process the thermal conductivity of the dry preform is four times lower. This is visible in the slow heating of the temperature curves of the spars during the first heating phase from RT to 104°C. By reaching the injection point resin properties are added and thermal conductivity, heat capacity and density are changing (Figure 100 to Figure 102). After this point the thermal conductivity is a compound of resin and fibres. The heating from injection temperature to curing temperature is more uniform. It has to be noted that the injection point can be used to reduce the first heating phase significantly, but it has to be proven whether influence of curing on the viscosity is small to fill the whole cavity. In the following pictures, temperatures in the mould, in the mandrels and on the composite parts are presented on point 3 (time 280min).

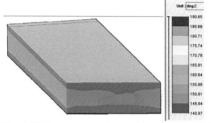

Figure 90 Temperature in the mould on a cut located in the middle of the mould (max. 190°C, min. 140°C)

Figure 91 Temperature in the mandrels (max. 175°C, min. 154°C)

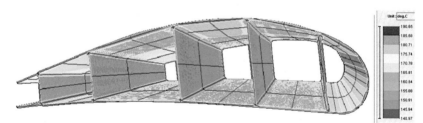

Figure 92 Temperature in the composite part (max. 190°C, min. 140°C), Time 280min, after reaching 180°C

The simulation shows that the temperature distribution in the mould is not homogeneous. During the heating from injection temperature to curing temperature the temperature of the composite part varied largely. The skin parts are heated fast, but the mandrels slow done the heating process and consequently the heating of the spars is delayed. The temperature has an extensive influence on the progress of curing (Figure 93).

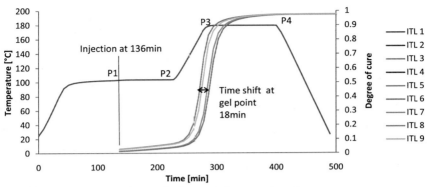

Figure 93 Press Temperature and curing development in the composite part

The development of the degree of cure is shown for different points in the upper skin, lower skin and in the spars. It is clearly visible that the curing is not homogeneous and the curing is sensitive to non-uniform temperature fields. Figure 93 can be taken to identify the first transition point, the gel point, from a liquid to rubber state. The gel point was taken from the experimental characterisation of chapter 3 with a value of 0.4. In the process, the gel point is shifted with a time delay of 18min. In Figure 94 the development of the glass transition temperature is shown.

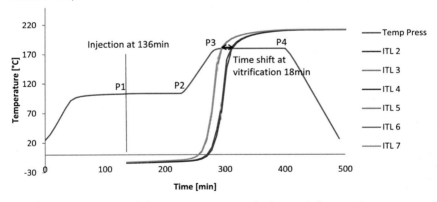

Figure 94 Press Temperature and glass transition temperature development in the composite part

The glass transition temperature is coupled using the DiBenedetto equation (eq. 3.7) to the curing reaction. The Figure 94 can be used to identify the next transition point, the so called vitrification point, when the material transforms from the rubber to a solid phase. The vitrification point occurs if the glass transition temperature is higher than the local temperature. Therefore this point is a process-dependent value. The time shift in the vitrification point is about 18min. The following figure 95 to 98 present the degree of cure at discrete time steps over the full structure.

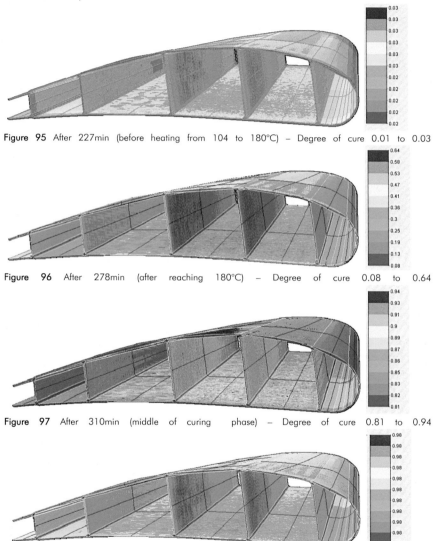

Figure 95 After 227min (before heating from 104 to 180°C) – Degree of cure 0.01 to 0.03

Figure 96 After 278min (after reaching 180°C) – Degree of cure 0.08 to 0.64

Figure 97 After 310min (middle of curing phase) – Degree of cure 0.81 to 0.94

Figure 98 After 400min (after cooling) – Degree of cure 0.98 to 0.98

Analysing the evolution of curing, it is visible that the curing starts before reaching the curing temperature of 180°C in the skin region. The region of the spars will be cured at the end. The curing is very inhomogeneous and dependent on the non-uniform temperature distribution amplified strongly by the autocatalytic reaction.

In comparison to the simulation, the degree of cure of the prototype no. 2 component is evaluated using the DSC method. The measurements are taken on different arbitrary chosen positions and are presented in the Tcble 7.1

Table 7.1

Sample name	Position	Degree of cure	Std. Deviation
CMF02-2000-3	Spar 3	97.87%	0.49%
CMF02-2100-LE	Leading edge	96.95%	0.57%
CMF02-2100-23-U	Lower skin	98.60%	0.10%
CMF02-2200-2	Spar 2	98.03%	0.43%
CMF02-2330-23-D	Upper skin	97.98%	0.70%
CMF02-5730-3	Spar 3	98.42%	0.06%
CMF02-5740-2	Spar 2	98.13%	0.26%
CMF02-6150-23-D	Upper skin	97.86%	0.23%

In comparisons between experimental evaluations of the degree of cure to the simulation results, a good accordance is found with an average deviation of less than 1% .

The material model is capable of providing a virtual material characterisation on the ply level. In the following these results are presented. The topology of the mechanical FE - model is built on the "As-Built" configuration which takes into account different discrete distribution of fibre volume contents. The distribution of fibre volume content values is based on variations of the cavity and has been measured by the skin thickness. These changes are of course, continuous. In the FE - model they are implemented as discrete values between the spars and by 8 sections from inboard to outboard sides using average values. The following Figure 99 presents this distribution, which is based on a measurement of the layup thickness.

Figure 99 Resulting fibre volume content, variation from 0.63 to 0.46

The "As-Planned" fibre volume content is around 0.55. The variation of the fibre volume content starts with a minimum value of 0.43 at the leading edge and a maximum value of 0.63 on the lower skin side. These variations have a significant influence on the resulting material properties and lead to non-uniform distribution of material parameters in the structure. Figure 100 displays the changes of the thermal conductivity over process time.

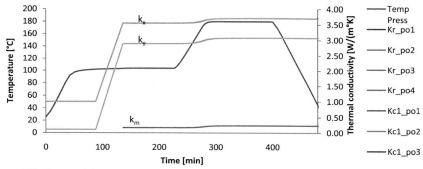

Figure 100 Changes of the thermal conductivity of the resin (k_m) and laminate in X (k_x) and Y (k_y) direction

In Figure 100, the changes of the thermal conductivity of the resin and the homogenised laminate values are shown. For the laminate, the homogenisations are presented in the direction of the global coordinates, X and Y. The details observed in chapter 3 and 4 the main changes of the thermal conductivity are influenced by the injection and the temperature. The changes in based on the degree of cure are less especially on the laminate level. The resulting cured values of the laminate are 3.70 W/m² in the X-direction and 3.06 in Y-direction. The following Figure 101 presents the changes of the heat capacity of the resin and the homogenised laminate.

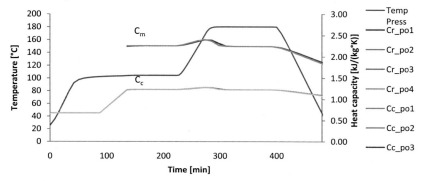

Figure 101 Changes of the heat capacity of the resin (c_m) and laminate (c_c)

A significant change of the value is driven by the injection of the resin, and after adding the resin, the changes dependent on temperature and curing reaction are nearly equal in the rubber phase. The cured material dependents much more sensitively on the temperature than the non-cured material in the rubber phase. To complete all thermochemical available results the density is presented in Figure 102.

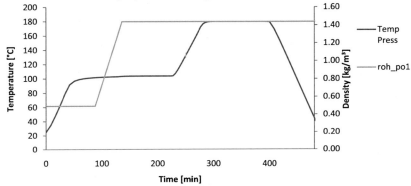

Figure 102 Changes of the homogenised density over process time

The following table 7.2 summarises the final cured thermochemical properties which are the result of the thermodynamical analysis.

Table 7.2	k_x	k_y	k_m	c_c	c_m	ρ_c	T_g	p
	[W/(m°K)]	[W/(m°K)]	[W/(m°K)]	[kJ/(kg°K)]	[kJ/(kg°K)]	[kg/m³]	[°C]	
	3.70	3.06	0.23	1.06	1.77	1.44	211	0.98

Additionally to the thermo dynamical results, the results of the mechanical analysis are presented. Figure 103 presents the changes of the modulus in fibre direction for some arbitrary chosen points over the process time.

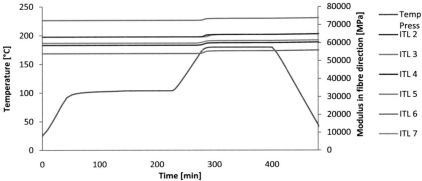

Figure 103 Changes and range of modulus in fibre direction over process time

The cured values of the modulus in the fibre direction varied from 73900 to 54930 MPa. The influence of the curing and the temperature is less in the fibre direction. The most significant influence is based on the fibre volume content. The following Figure 104 displays the variation of the fibre modulus over the full structure.

Figure 104 Distribution of the modulus in fibre direction after the curing process [MPa]

It is visible that the discrete distribution is an approximation. For more reliable results, continuous mapping methods of experimentally achieved data such as thickness, fibre volume content would be helpful to improve accuracy. Figure 105 presents the changes of the resin modulus.

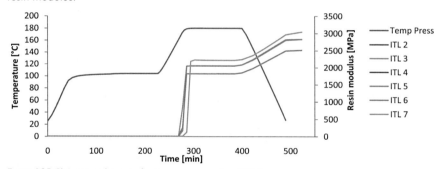

Figure 105 Changes and range of resin over process time [MPa]

The influence of the resin is dominant in the transverse direction. Accordingly, over the process time the curing start in the upper skin. Figure 106 shows the modulus in the transverse direction at time step 286min. The values varied from 19 (rubber phase) to 7940MPa (solid phase).

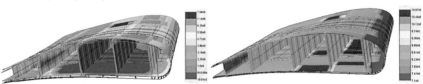

Figure 106 Modulus in transverse direction during curing at 286 min – Range 7940/19MPa – and at the process end – Range 10870/7100MPa

113

On the right side of Figure 106 the cured modulus is shown. The variation is between 7100 to 10870 MPa. The table 7.3 summarises the available material values and their minimum and maximum value.

Table 7.3	E_1 [MPa]	$E_2=E_3$ [MPa]	$G_{12}=G_{13}$ [MPa]	G_{23} [MPa]	ν_{12}	ν_{23}	α_1 [/°C]	$\alpha_2=\alpha_3$ [/°C]
Max.	73900	10870	4470	3820	0.28	0.46	0.97E-06	32.6E-06
Min.	54930	7100	2790	2460	0.26	0.46	0.29E-06	26.3E-06

The interpretation of the results of the resulting engineering properties leads to the conclusion that the fibre volume content is significant influence. The variation of the resulting properties is high and especially variation of the CTE is critical. The following

Figure 107 107 and Figure 108 indicate the distribution of the coefficient of thermal expansion CTE1 and CTE2 over the full structure.

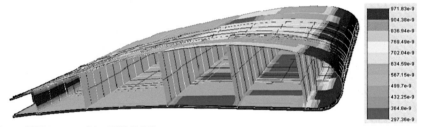

Figure 107 Variation of the CTE1 [1/°C]

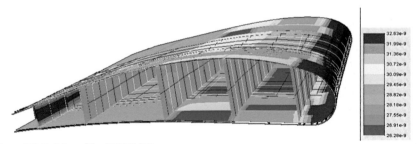

Figure 108 Variation of the CTE2 [1/°C]

The variation of the CTE from the upper skin to the lower skin results in a disadvantageous situation which influences the process-induced deformation of this box structure significant. In the following Figure 109 the process-induced deformations are presented.

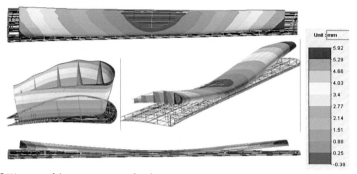

Figure 109 Warpage of the composite part [mm]

The displacement is displayed as scalar magnitude projected in out-of-plane direction, perpendicular to the flap. This representation of the process-induced deformations results in a displacement from -0.36mm to 5.92mm. This means a total magnitude of 6.3mm. In Figure 110 110, the experimental measured deformations of prototype nr. 7 are shown. The measurement was undertaken using a structured light 3-D scanner, namely the Atos system from GOM. The analysis of the experimental data was performed using software GOM Inspect which compares the measurement with a CAD Reference. The problem of a measurement is always to find the same reference coordinates. In the case of a deformed box structure, as in the given case this is difficult. Accordingly, there are some methods like "best fit" for the alignment of the measured data to the CAD. This method was used in this application, because no reference coordinate system was defined previously. The representation of deviation values is different to the FE representation. In FE software the displacement values are always based on a coordinate system. In the case of the GOM Inspect software, the deviations to the CAD are calculated in the normal direction to the surface. Therefore, the absolute values are not comparable. Figure 110 shows the deviations of the upper and the lower sides.

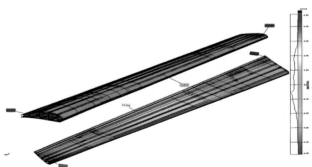

Figure 110 Measured warpage of the composite part [mm]

115

The measured warpage from the upper side shows a minimum value of -5-71 and a maximum value of 2.55mm, which lead to a total deformation of 8.26mm. The lower side deforms from a minimum of -4.56mm to 2.40mm which leads to a total deformations of 5.96mm. This measurement was repeated in the CMF project with a laser tracking system to have a comparable evaluation of the measurement method. The result of this measurement was a magnitude of 5.96mm. This highlights a general problem of the analysis of warpage of large composite structures. A reliable experimental basis for the validation of numerical methods, especially for large flexible structures like a thin fuselage skin or a panel are not self evident. The definition of reference coordinate systems, the alignment of experimental data and the possibility to compare equal values like out-of-plane displacement or deviation of the normal face direction lead to a lack of imprecision. Nevertheless the numerical results show the same deformation behaviour, the curved banana shape, and a magnitude of deformation which is between the GOM measurement and the laser tracking method.

In Figure 111 the material effort is presented over the process time. To condense the complex stress state, the Puck inter fibre effort value is used to reduce the stress tensor to one scalar value. The positions of the points are arbitrarily chosen.

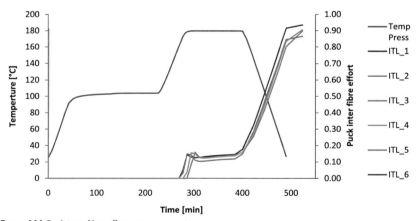

Figure 111 Puck inter fibre effort over process time

The development of material effort starts after reaching the gel point. It increases strongly in the rubber phase, based on the chemically induced shrinkage. The material effort based on chemically induced shrinkage is around 15 to 20%. After this, due to the viscoelastic relaxation effect, the effort decays during the isothermal curing. In the cool down phase the thermal induced shrinkage increases the material effort up to a value of 0.85 to 0.98. Figure 112 presents the distribution of the material effort value over the structure. It is clearly visible that in regions of high fibre volume content, the stiffness is higher and, consequently, also the material effort values are higher.

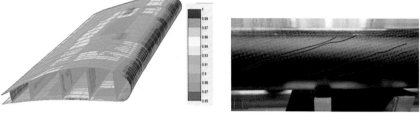

Figure 112 Puck inter fibre effort   Figure 113 Visible surface cracks

The calculated material effort is close to the failure index of 1. This means that inter fibre failure will occur. The application of the Puck criterion was performend using the cured values of the material from the datasheet for G1157. The used material is a fabric, namely the G0926. Therefore this is an approximation of the effort in this case. One validation method for this value can be performed by comparing the optical inspection of the quality of the finished part. In Figure. 113, some visible surface cracks could be detected on the leading edge. Consequently the calculated values are in a realistic range. In Figure 114 shows the stress distribution transverse to the fibre direction.

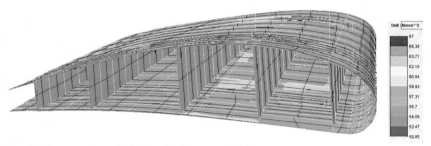

Figure 114 Transverse stress distribution after the process [MPa]

7.3 Sensitivity of influencing factors

The table 7.5 presents all parameters which will have an influence on the warpage of the CMF and their measured variations during the development phase of the first 9 prototypes. The parameters can be classified into internal topology parameters (layup errors, wrinkling, orientation errors, voids, etc.) global topology parameters (tool inaccuracy, wrong mounted mandrels, etc.), process parameters (variation of heating rate, curing temperature, etc.) and shape parameters. All the listed parameters affect on the warpage by influencing the coefficient of thermal expansion (CTE), the temperature delta (ΔT) or on the resulting lengths (l_0).

Table 7.5 Parameter	Variation	Sensitivity	Influence
Undulations	no	small	CTE
Wrinkling	1 detected	small	CTE
Pre form – layup errors	fibre volume content 46-72%	large	CTE
Voids	no	small	CTE
Fibre volume gradients	Est. 10%	middle	CTE
Tool inaccuracy	Thickness, fibre volume content	large	CTE
Deformation of mandrels	Thickness, fibre volume content	large	CTE
Thermal expansion mould	Thickness, fibre volume content	large	CTE
Variation of curing temp.	2-3°C	small	ΔT
Variation heating rate	1.52-1.92°C, inhom. curing	middle	CTE
Variation cooling rate	unknown	--	---
Variation of cure time	106-116 min	small	CTE
Variation of holding time	173- 575min	small	CTE
Unsymmetrical shape	628 -654mm	large	l_0

In the following study, the "As-Built" model is used to perform parameter studies to show the sensitivity and variability of the process with respect to the parameters of thickness variation, heat rate and weft / warp influence on the layup. As shown in the table 7.4 these parameters have been identified by a review of the experimental data as major parameters influencing the process-induced deformations. Figure115 to Figure 118 presents the results of single simulation runs. In the model, shown in Figure 115, the thickness variation was neglected. This means that there is no variation of the fibre volume content and no variation of the CTE from the lower to upper side of the flap.

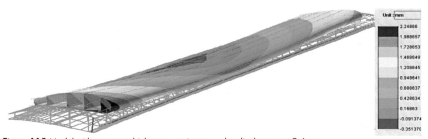

Figure 115 Model with constant thickness, maximum scalar displacement 2.6mm

In the following two models only the heat rate was modified. The "As-Built" model has a second heat rate (from 104 to 180°C) of 1.5 °C/min. In the first model a heat rate of 1°C/min and in the second run a heat rate of 2°C/min was applied.

Figure 116 Model with heat rate 1K/min, maximum scalar displacement 5.62mm

Figure 117 Model with heat rate 2K/min, maximum scalar displacement 6.34mm

In the fourth parameter study the skin layup was changed from [45/-45,0/90,90/0,45/-45,0/90,90/0,-45/45] to the following [45/-45,0/90,90/0,45/-45,0/90,90/0,45/-45]

Figure 118 Model with changed weft/warp directions, maximum scalar displacement 6.59mm

The following table 7.6 summarise the results of the different studies. As first result it can be observed that the deformation mode is not influenced.

Table 7.6	Max out of plane deformation	Variation to ref.
Ref. model based on LSD7 (As Built confg.) With variable thickness Heat rate 1.5 K/min	6.29 mm	
Constant thickness	2.60 mm	58.7%
Heat rate 1 K/min	5.62 mm	10.7%
Heat rate 2 K/min	6.34 mm	-0.80%
Layup changes (non-flipped)	6.59 mm	4.77%

119

The standard deviation of the numerical method based on the material input parameter is about 2%. This variation is based on the variation of the material input parameters and the accuracy of determining the input values (chapter 8). The model standard deviation was determined using a Monte Carlo Simulation which takes the standard deviation of each material input parameter into account. For the total standard deviation of the model the uncertainties in the description of the process have to be added also. This means variation of process parameters like curing temperature, heat rate e.t.c and they are listed in table 7.5. To estimate the total standard deviation of the model, 6% of general uncertainties have to be added to a total accuracy value of 8%.

Analyzing the results of the parameter studies this means that lay up variation based on weft/warp directions, temperature variation and the influence of the heat rate are second order effects which create a variability of the process about 10%. The main driver of process-induced deformations is the variation of thickness. The variation of the thickness driven by displacement of the mandrels changes the fibre volume content and consequently the coefficient of thermal expansion. The difference between the "As-Built" configuration and the result with constant thickness is about 59%. Accordingly this is the main influence parameter for the warpage of the composite multispar flap.

The mould tolerance of the thickness is about 0.2mm. This variation has a large impact on the fibre volume content and, on the resulting coefficient of thermal expansion. In the case of a thickness deviation of the lower skin to the upper skin, this influence is the key factor for process-induced deformations. The variation of the thickness can have several causes like wrong tool geometries, mounting errors of the mandrels, deformation of the mandrels (based on thermal expansion and injection pressure gradients). The second most sensitive parameter on the distortion is the unsymmetrical shape of the box structure. The third main parameter is the heating rate, which leads to an inhomogeneous temperature / curing condition. Therefore, the main driver which affects the resulting process-induced deformations can be summarised as follows:

1. Thickness deviation
2. Unsymmetrical shape
3. Heating rate

The parameters can be again classified into controllable and uncontrollable parameters. The outer shape of the CMF cannot be modified, therefore, it is of an uncontrollable type. The thickness deviation is based on the global tolerance of the mould ($\pm$ 0.1mm manufacturing tolerances of all surfaces) and errors. Hence, this parameter can be only minimally controlled and is also of the uncontrollable type. The heating rate can be classified as a controllable parameter. In the following section, some ideas about warpage compensation are presented and discussed.

7.4 Warpage compensation strategies

In general, two major strategies can be applied to minimise the resulting process-induced deformations. First of all, the geometry of the mould can be modified to compensate the warpage. This process is quite complex, time consuming and expensive for large integral composite parts. The rework or redesign of a mould can be based on simulation results or experience. Both sources of knowledge are more or less reliable and, in the worst scenario the redesign of the mould, is a kind of iterative process. The second possibility is to influence the warpage by the process parameters directly without changing the topology of the mould and the part. As shown in the previous study, the curing process is not homogeneous due to the complex shape of the CMF component. Therefore, this situation can be used by applying different curing temperature conditions to compensate the total amount of distortions. The simulation method is a suitable method to evaluate the process parameters virtually with fewer amounts of costs. The used heating press has the opportunity to apply different temperature boundary conditions by 12 heating fields on the lower and upper side of the heating press.

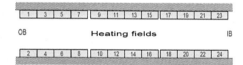

Figure 119 Heating press and heat fields

The deformed shape of the CMF component shows two types of deformation mode: a bending (banana) mode and a small twisting mode. The main mode is the bending deformation and this can be influenced by a temperature gradient of the upper to the lower side of the heating press. Figure 120 illustrates the results of two different simulation runs with a temperature gradient of 20°C of the lower and upper side of the heating press.

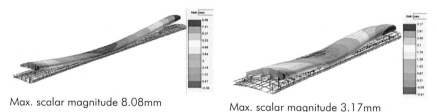

Max. scalar magnitude 8.08mm

Max. scalar magnitude 3.17mm

Figure 120 Warpage of the component with a temperature gradient of 20°C (left - lower side 190°C/upper side 170°C; right – upper side 190°C/lower side 170°C)

The result of this sensitivity analysis shows that, in general, temperature boundary conditions can be used for warpage compensation. The sensitivity of the temperature boundary conditions is high and changes the warpage behaviour. The resulting final degree of cure, which is shown in Figure 121 is influenced by less than 1.5%.

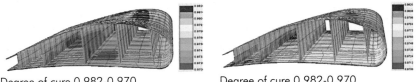

Degree of cure 0.982-0.970   Degree of cure 0.982-0.970

Figure 121 Final degree of cure of the component with a temperature gradient of 20°C (left - lower side 190°C/upper side 170°C; right – upper side 190°C/lower side 170°C)

The industrial usage of this effect might be interesting, but existing manufacturing restrictions such as temperature variation should be lower than 10°C and will not be fulfilled. Another aspect is that thermal deformations of the mould are different to the initial geometry and can also influence the part geometries, especially if aluminium as a tool material is used. Nevertheless, it is possible to influence the warpage by the temperature, significantly.

7.5 Conclusion

The developed simulation strategy and material model have been successful used to analyse the manufacturing process of the composite multispar flap and it was possible to achieve reliable results. One of the main problems during the analysis of the flap was discovered between the so-called "As-Planned" configuration and the "As-Built" configuration. It was found that modelling of a complex problem like this should always be in close cooperation with the real process, because during the manufacturing a permanent uncertainty exists, and the process is very sensitive to, for example, layup, preform errors, temperature changes, etc. Therefore, the model is suitable to analyse the process, but it needs also a high amount of experience to achieve the required qualities.

The following results were found:
- The highest uncertainty of the model exists between the "As-Planned" and "As-Built" configuration. This means that, for example, if preform errors are not implemented into the FE model the results will be erroneous. Consequently, the quality management system which should document the process must be highly sophisticated.
- Improved measurement methods are needed to close the gap between virtual and experimental analyses, especially for the measurement of the final shape of the part.
- The simulation strategy and the material model achieve reliable results and can reproduce the physical behaviour of the component in the process.
- The model is suitable to perform parameter studies on risk parameters to achieve a higher process understanding.
- The model is convenient for virtual studies of warpage compensation.

8 Further application based on non deterministic methods

Chapter 8 is presenting an outlook on the application of the analysis method with respect to intelligent processing tools for composite. There are different methods for planning process conditions. In this chapter, first an overview about existing methods is presented. In a second part, some examples are given to couple the developed simulation strategy to non deterministic methods.

In the current development of a manufacturing process, a high effort is implied to find the optimum of process parameters to achieve the required quality. Most of the time, a trial-and-error method is used, which is sometimes a random process. The development process starts, usually, with a parameter set which is based on the manufacturer and on material provider specifications. Different variations are then tried out with the process parameters until an acceptable quality is reached or time / funding are depleted. The best parameter sets are then chosen. Therefore, very conservative process parameters are selected and the tolerances of the parameters are very high to ensure a repeatable process with qualities inside the so-called "process window". The main advantage of such a trial-and-error method is that it does not require a high level of knowledge about the process itself. The disadvantage is that it is not practical to perform all possible variations in order to find the real optimum of the process, because a lot of mechanical and physical testing is needed to determine the product quality. Therefore, it is difficult to transfer the process knowledge to new materials, processes or parts. Also, there is no information about the perturbation of the process in view of the variability to material or process parameters.

In the last time there goes a trend to use non deterministic methods to evaluate the sensitivity of process parameters and the variability of the process. Three methods are presented to reduce the number of experimental trials and to acquire, at the same time, more information about the process itself. The first method is based on a Design of Experiment (DOE) to evaluate the sensitivity of process parameters. The second method is based on Monte Carlo Simulation (MCS). In a third step, an optimisation method is demonstrated to find optimum process parameters.

In the following chapter, both probabilistic methods and the optimisation approach are applied to analyse the experimental and numerical model of validation test case 2. This simple plate with the layup of $[0_3/90_3]$ will be used for the sensitivity analysis, variability analysis and application of process optimisation method. All studies are performed using the multidisciplinary optimisation platform Boss Quattro.

8.1 Sensitivity analysis

One of the most important applications of a process simulation model is the possibility to perform sensitivity analyses to identify major factors on the process variability. This is a key point to reach the quality level and to minimise the effort of quality control on process input parameters. The identification of the material, process and geometrical parameters which effect on spring-in, warpage and residual stress can be down using different methods. One method is to perform parameter studies by variation of one single parameter and analysing the response of a system. This method is fast, but limited if the system response is not solely one value. For example, in a composite part, different process-induced deformations appear (spring-in and warpage). To analyse the impact and the dependency, a more advanced method can be used, such as "Design of Experiment" (DOE). The advantage of a DOE is that, also the dependency between different influencing factors and their impact can be analysed [81, 4]

In general, DOE is a phenomenological approach. An empirical or mathematical model can be observed by analysing the response of the system in performing a variation of the influential factors. These factors can have the nature of quantitative (continuous variables) or qualitative (discrete variables) types. The response of the system is analysed using multiple linear regression methods by creating a response surface. This answers the question about the influence of each factor on the response, measures the coupling between factors and gives an account for the linear or nonlinear influence of certain factors. In the case of a phenomenon subjected to the influence of two parameters A and B, the traditional approach consists of studying separately the two variables, making A vary from A to A for a mean value of B, and emulating this for B. This does not allow for creating a response surface over the full design domain. To achieve this design, the domain should be meshed and experiments should be performed at certain nodes. This method is computationally costly, especially for a system parameterised with, for example, seven parameters. Using a nonlinear fashion with 3 points, at least 2187 tests should be performed. Different types of designs have been developed to reduce this large amount of experiments. One restriction is to define orthogonality, which means factors A and B are independent, but A and AB are not independent. Another method is to use full factorial designs techniques to reduce the number of tests.

8.1.1   Application 1 - Design of Experiment on material parameters

In general, parameters can be categorised into material and process parameters. The following DOE will be limited to mechanical analysis part and on material parameters. The aim is to prove the sensitivity of the process as opposed to the variability of the characterised material parameters. The characterisation of the material is always affected by variation, which can be determined by the standard deviation of the measured value. In chapters 3-4, all important material values have been measured, including these variations. The objective of this study is to identify the sensitivity of the single material parameters, based on their variations. By using the developed material model 23 parameters (without curing) are used to characterise the mechanical behaviour. This leads to a high number of virtual experiments. Therefore, it is useful to make a selection of the most important parameters. From the literature review, Svanberg [21] mentions that the most relevant parameters on distortions are thermal expansion, shrinkage and fibre volume content. Therefore, the following values in table 8.1 are chosen for the DOE:

Table 8.1

Parameter	Symbol	Abbr	Value	Deviation %	Max. Value	Min. Value
cured resin modulus	$E_{me}\ [N/mm^2]$	ERE	2890	6.5	3078	2702
cured resin Poisson ratio	ν_m	POIM	0.34	2	0.35	0.33
fibre modulus	$E_{f1}\ [N/mm^2]$	EF1	210000	2	214200	205800
fibre shear modulus	$G_{f12}\ [N/mm^2]$	GF12	50000	2	51000	49000
fibre Poisson ratio	ν_{f12}	POIF	0.23	3	0.23	0.22
fibre volume content	V_f	PHI	0.63	5	0.66	0.59
resin chemical shrinkage	γ_m	SCH	0.34	5	0.36	0.32
fibre CTE 1	$\alpha_{f1}[1/^\circ K]$	ALF1	-8.00E-07	10	-8.80E-07	-7.20E-07
fibre CTE 2	$\alpha_{f2}[1/^\circ K]$	ALF2	5.50E-06	5	5.78E-06	5.23E-06
resin CTE	$\alpha_m[1/^\circ K]$	ALM	5.00E-05	5	5.25E-05	4.75E-05

A variation of these material factors, based on the standard deviation from the measurement of chapter 3-4, is applied. A 2-level factorial design is used, which lead to a design table with 1024 analyses. For the response of the analysis the maximum out of plane displacement of one of the corner nodes at the last time-step is chosen. Figure 122 displays the sensitivity of the displacement, without all interactions of the parameters.

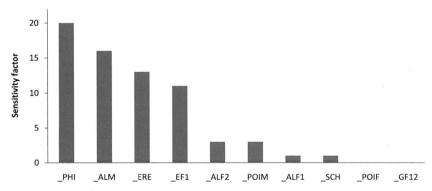

Figure 122 Sensitivity of material parameters on warpage

DOE results that the most significant parameter on the quality of the result is the fibre volume content (_PHI), the coefficient of thermal expansion of the resin (_ALM) and the resin modulus (_ERE). Unfortunately, the fibre volume content is not a real material parameter, because it is based on topology and / or process conditions. The fibre volume also changes during the process by the influence of the matrix shrinkage and it is possible to have a variation of the fibre volume content over the thickness. The fibre volume content is afflicted with a high degree of uncertainty and this leads to a disadvantageous situation, because most of the resulting material parameters are based on the fibre volume content. The accuracy of the material model is strongly coupled to the accuracy of the fibre volume content parameter. The following disturbing qualities can affect the fibre volume content:

1. Variation of mould dimension, especially variation of cavity thickness
2. Preform/Layup errors
3. Draping errors, especially in corners
4. Variation of fibre volume content over the thickness by different degrees of compaction
5. Change of initial fibre volume content by chemical and thermal shrinkage

To design a robust, efficient manufacturing process these disturbance quantities are key parameters.

The second results of the performed DOE are the sensitivities of the parameters on the material effort. Similarly to the previous chapter, the Puck inter fibre effort value is used. Figure 123 shows the sensitivity of the material parameters on the inter fibre material effort.

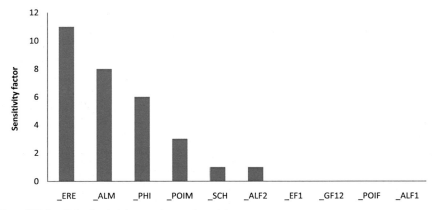

Figure 123 Sensitivity of material parameters on stress state

In this case, the resin modulus (_ERE) and the coefficient of thermal expansion (_ALM) of the resin are the most sensitive parameters on the stress state. Note, that this is different to the results for the sensitivity on the process-induced deformations.

It has to be mentioned that this sensitivity analysis does not show the sensitivity of the process in general. It is specific to the manufacturing process (RTM), the used materials and the accuracy of the material characterisation. One result of this analysis is the reliability of the numerical model in coherence to the uncertainties of the parameter measurements. The accordance of the experimental measured parameters with the highest uncertainty (fibre volume content, coefficient of thermal expansion CTE1) and the most sensitive parameters of the numerical model (fibre volume content, coefficient of thermal expansion – matrix) is critical with respect to the trustability of the numerical result. It identifies that advanced methods are needed to measure these parameters with a higher precision and identifies them as critical parameters.

8.1.2   Application 2 - Design of Experiment on process parameters

The second application of the DOE is performed on the process parameters. Again, the model of validation test case 2 is used to identify the most significant process parameters. The following process parameters can be identified from Figure 72:

Table 8.2 Parameter	Unit	Abbr.	Value	Max. Value	Min. Value
Heat rate	$[°K/min]$	AURAT	1.07	1.43	0.94
Cooling rate	$[°K/min]$	ABRAT	1.49	1.3	2.0
Curing time	$[min]$	HAL	90.00	120	80
Curing temperature	$[°K]$	TEMP2	453.00	462.06	443.94
Room temperature	$[°K]$	TEMP3	293.00	298.86	287.14
Fibre angle ply 1	$[°]$	W1	90.00	93.0	87.0
Fibre angle ply 6	$[°]$	W6	0.00	3.00	-3.00
Fibre volume content	V_f	PHI	0.63	0.66	0.59

This table 8.2 is based on parameters which can be changed by the manufacturing designer based on the specifications of the process. For some of these parameters such as heat rate, cooling rate and curing time restriction emanate from the material provider exists. One of these limitations, for example, defines the max. heat rate to 5°K/min or limits the variation of the curing temperature to ±5°C. The table shows the variation of the parameters which have been taken as arbitrary. To the real process parameters the fibre volume content was added as a dependent parameter for the autoclave/vacuum pressure. Also, two parameters for the fibre orientation were added. The fibre orientation is always based on the tolerance specification during performing and variants in these tolerance limits.

For the DOE, again a two-level factorial design is chosen which leads with the 8 parameters to a 256 point analysis. For the response of the analysis, the maximum out of plane displacement of one of the corner nodes at the last time step is chosen. Figure 124 shows the sensitivity of the single parameters without interaction.

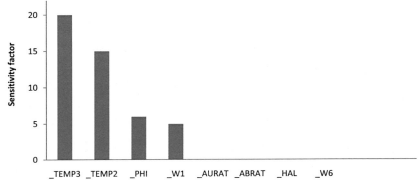

Figure 124 Sensitivity of process parameters on warpage

The sensitivity plot shows that temperature conditions have a large impact on the results. The variable TEMP3, the final temperature (room temperature), and TEMP2, the curing temperature are most sensitive. In view of the temperature dependent engineering constants, the influence of the final temperature is higher than the curing temperature. The high sensitivity of the temperature conditions is obvious, but it has to be taken into account that a lot of different processes exist with a non-uniform temperature field. This was demonstrated in the previous chapter by analysing the RTM process of the CMF component and this is also the case for a manufacturing process in the autoclave.

The third most sensitive parameter is again the fibre volume content. The fibre volume content was identified in the previous study as the most sensitive material parameter, but the nature of the parameter is also process-dependent. The fourth most sensitive parameter is the orientation of the 90° direction layer. In transverse direction to the fibre the coefficient of thermal expansion is very high. A small variation of this layer can change the value of the ply significant and decreases the process-induced deformations. The other process parameters such as the heat rate, the cooling rate, the orientation of the 0° direction and the curing time do not have a significant influence on the process-induced deformations. In Figure 125 the sensitivity of the process parameters on the stress state is shown.

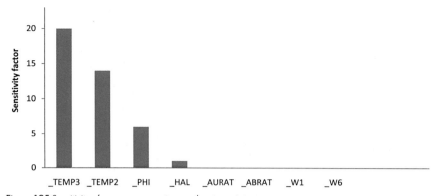

Figure 125 Sensitivity of process parameters on the stress state

Similar dominating parameters can be identified in this result of the study for the first three parameters which affect on the stress state. The position of the fourth parameter, HAL the curing time, is different. At a high curing time the chemical-induced shrinkage stress components can relax and, therefore, a long curing time can decrease the material effort.

It has to be mentioned that these two studies are an example of a sensitivity analysis based on the experimental data of validation test 2. The variations of the material and process parameters are derived from measurement variations and process parameter variation. The sensitivity of the parameters is specific to the problem and the results should be considered carefully in order to generalise them to other processes, layups and parts. Nevertheless, the method "Design of Experiment" is demonstrated and is suitable to derive the most sensitive process parameter: the temperature, the fibre volume content and the fibre orientation.

8.2 Variability analysis

During a manufacturing process a lack of knowledge exists concerning the variation of part quality in respect to variability of raw material and process parameters. The variability of a composite manufacturing process makes it difficult for the manufacturer to control and adjust the process and leads to rejections or additional rework.

Therefore, variability represents a random change of these parameters, maybe between specific tolerances. The response of a system, for example a production line, will be the production quality which is between the specifications, or not. The manufacturing of composite parts for aerospace application has a high demand on quality with only small variations in the quality. As shown in previous chapters, the manufacturing process is based on many different parameters, because the material properties are created by the process itself. The manufacturer has to choose between the variability of the material input by defining tolerances for specific parameters. In general, if the manufacturer selects a material with low quality, this will lead to higher assembly or rework cost. This leads to a kind of optimisation problem which is shown in Figure 126 [82].

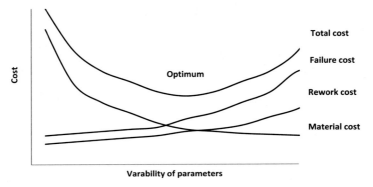

Figure 126 Schematica overview of dependency of cost and variability [82]

Therefore, methods are needed to prove the influence of specific tolerances on the variability of the production quality in order to develop cost-efficient manufacturing methods. This study presents a Monte Carlo variability analysis applied on material and process parameters. The Monte Carlo method was developed in the 1940s and is used to analyse systems with significant uncertainties. For each chosen parameter of the analysis a certain randomly variation which is based on Gauss distributions within the specific range of the parameters is assigned. For the given example, the numerical model of validation test case 2 is taken and the parameters, variations and number of variation steps are displayed in the table 8.3. The numbers of steps are taken arbitrary and lead to 330 analyses.

Table 8.3 Parameter	Units	Abbr.	Value	St. Dev. [%]	Nr. Steps
cured resin modulus	E_{me} [N/mm^2]	ERE	2890	6.50	30
fibre modulus	E_{f1} [N/mm^2]	EF1	210000	6.50	30
fibre volume content	V_f	PHI	0.68	15.0	30
resin chemical shrinkage	γ_m	SCH	0.34	10.0	30
fibre CTE 1	α_{f1}[1/°K]	ALF1	-8.00E-07	5.00	30
fibre CTE 2	α_{f2}[1/°K]	ALF2	5.50E-06	1.00	30
resin CTE	α_m[1/°K]	ALM	5.00E-05	5.00	30
Heat rate	[°K/min]	AURAT	2	2.50	20
Cooling rate	[°K/min]	ABRAT	1.49	1.30	20
Curing time	[min]	HAL	90	5.00	20
Curing temperature	[°K]	TEMP2	453	5.00	20
Fibre angle ply 1	[°]	W1	90.00	5.00	20
Fibre angle ply 6	[°]	W6	0.00	5.00	20

Table 8.4 and 8.5 present the variation of distortions and stresses displayed by the effort value for inter fibre failure. The results of the Monte Carlo analysis are the average value, the standard deviation, the average deviation and the variance of the process.

Table 8.4 Inter fibre effort

Mean	0.51
St. Dev	0.006
Average Dev.	0.002
Variance	$3.8e - 5$

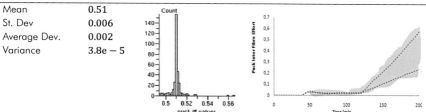

Table 8.5 Deformation

Mean	37.68mm
St. Dev	0.47mm
Average Dev.	0.23mm
Variance	0.22 mm

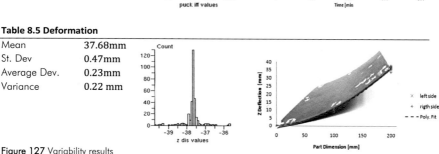

Figure 127 Variability results

For the validation test 2, a total of 8 plates with the same process conditions have been cured. The natural deviation of the process was determined with a standard deviation of 1.9mm of the Z deflection [S4]. The result of the Monte Carlo simulation results in a standard deviation value of 0.47mm which is four times to low. The chosen input parameters are arbitrary. They are material and process parameters. The variability of material parameters have been taken from the experimental characterisation chapter 3-4 using the standard deviation. The process parameters variability has been estimated. Especially the fibre volume content, which is one of the most sensitive parameters, was taken with a standard deviation of around 15%. There are further investigations needed to improve this method to be reliable.

The example of this study demonstrates the Monte Carlo Simulation to evaluate the variability of the process. A process designer can achieve, with this method, the tolerances of the process parameters to get the required tolerances and qualities in the final product.

8.3 Process optimisation

As shown in this thesis, the manufacturing process of composite is a multiscale, multiphysical problem and especially if the manufacturing costs are also taken into account, another dimension has to be added. For this multidisciplinary problem the real process optimum is nearly impossible to find by the trial and error method. Therefore, more sophisticated methods have to be applied in advance to achieve a stable, cost efficient, improved process. One solution might be to apply an optimisation method. To demonstrate such a kind of optimisation, the model and the process of the validation test case 2, is taken. In general, an optimisation problem has to be defined by an objective function, some design variables and some design constraints. In this case, the objective function is arbitrarily taken to decrease the process dependent deformations and the Puck inter fibre effort value at the same time. It is also possible to add more design functions as demonstrated in the previous chapter. As design variables, the process parameters are taken in respect to the bounds of the material provider specification. The following design variables are chosen.

Table 8.6 Design Variables	Units	Abbr.	Min. Value	Max. Value
Heat rate	$[°K/min]$	AURAT	0.01	5
Cooling rate	$[°K/min]$	ABRAT	0.01	20
Curing time	$[min]$	HAL	45	180
Curing temperature	$[°K]$	TEMP2	403	473

The optimisation was performed with the multidisciplinary optimisation platform Boss Quattro. The program BOSS Quattro provides some powerful optimisation algorithms, such as gradient methods (ConLin, SQP, GCM etc.) and genetic algorithms. In this case, the gradient-based algorithm GCM (Globally Convergent Method) was chosen, which is a multi-objective algorithm of second order. The optimisation algorithm is able to solve a non-monotonous function. The following functions (Figure 128 to Figure 129) present the variation of the objective function value of the number of iterations.

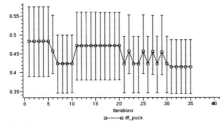
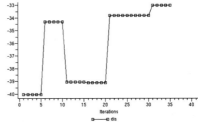

Figure 128 Variation of the Puck inter fibre effort value, result 0.49

Figure 129 Variation of the Z displacement, result 33mm

After 35 iterations, the optimisation algorithm found a solution with a minimum value of 0.49 for the Puck inter fibre effort value and a minimal deformation of 33mm. The following diagrams show the variation of the design variables:

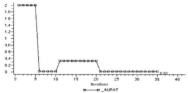

Figure 130 Variation of heating rate, result 0.01 K/min

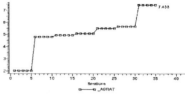

Figure 131 Variation of cooling rate, result 7,433 K/min

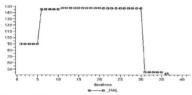

Figure 132 Variation of curing time, result 45 min

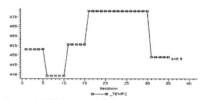

Figure 133 Variation of curing temperature, result 448 °K

The following optimal process parameters lead to a process which has a very low heating rate of 0.01°K/min.

Table 8.7 Results optimisation run no. 1

Parameters	Units	Abbr.	Optimal Value
Heat rate	$[°K/min]$	AURAT	0.01
Cooling rate	$[°K/min]$	ABRAT	7.43
Curing time	$[min]$	HAL	45
Curing temperature	$[°K]$	TEMP2	449
Puck Inter fibre effort		IFF	0.49
Z - displacement	$[mm]$	dis_z	33
Total process time	$[h]$	P5_t	94

The total time of the process is about 21h. This might be, from a scientific point of view, an interesting process condition, but it is not practical, because the cycle time is too long. Therefore, in the next optimisation run, the reduction of the total process time is added to the objective functions.

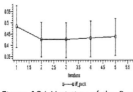

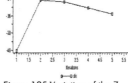

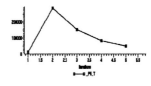

Figure 134 Variation of the Puck inter fibre effort value, result 0.51

Figure 135 Variation of the Z dis-placement, result 36mm

Figure 136 Variation of the total process time

The results of the second optimisation run show that by adding objectives to the problem, the GCM algorithm is able to compute the optimal set of parameters much faster, in only 5 Iterations. It is also visible that the achieved process time of 8.9h is a large reduction of the overall process time, but in the sense of cost efficient production is not in the frame of a cost efficient process. Therefore, if an optimisation like this should be performed, a coupling to the economic feasibility has to exist. This could be performed with weighted objective function or adding a real cost model to the optimisation problem definition. Table 8.8 summarises the evaluated parameters of optimisation run 2.

Table 8.8 Results optimisation run no. 2

Parameters	Units	Abbr.	Optimal Value
Heat rate	[°K/min]	AURAT	0.15
Cooling rate	[°K/min]	ABRAT	6.51
Curing time	[min]	HAL	92
Curing temperature	[°K]	TEMP2	455
Puck Inter fibre effort		IFF	0.51
Z - displacement	[mm]	dis_z	35.8
Total process time	[h]	P5_t	8.9

The following diagram displays the two different temperature conditions for an optimal process with less residual stress and the optimal conditions for a non-weighted residual and time minimum.

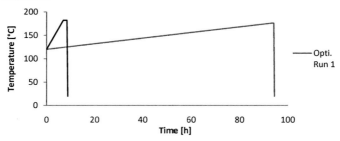

Figure 137 Optimal process conditions, Optimal Run1 – min residual stress, Optimal Run 2 - min time & residual stress

8.4 Conclusion

This chapter presented different nondeterministic methods in connection to the developed analysis method and material model. It was demonstrated that a coupling can be done to achieve more information about the sensitivity of process and material parameters, the influence of these parameters on the variability of the process and, at least, it was demonstrated how to evaluate optimal process parameters using a gradient-based optimisation algorithm. For further application, the following procedure could be usable:

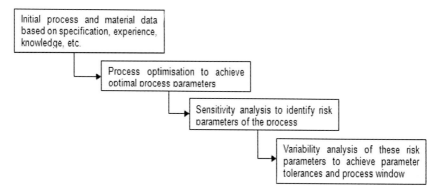

Figure 138 Process development process using nondeterministic methods

9 Summary

Process simulation, as one method to develop robust efficient manufacturing processes for composite parts, has become increasingly more important in the last decade. The possibility to perform certain kinds of analysis on risk parameters of the production process can help to accelerate process design time, improve product quality and reduce the number of deficient parts.
The focus of this thesis was to develop a virtual process chain for the RTM process to analyse process-induced deformations and stresses.

In chapters 3 and 4, the resin and composite behaviours were studied in detail, starting from the important physical phenomena which have been idealised by constitutive equations and implemented in a material model. It was found that the resin behaviour is relatively complex and depends on the temperature, the degree of cure and the glass transition temperature.

Later in chapter 5, a micro-mechanical study was performed in order to analyse the formation of residual stresses on this scale. As one result of the micro-mechanical study, it could be concluded that in combination with fibres in a composite, there is an interaction between residual stress, microplasticity and microdegradation. The following interdependencies were indentified:

- Analysing process-dependent residual stresses using temperature loads is not sufficient and leads to wrong maximum stress values and wrong stress distributions.
- Adding chemical shrinkage to thermal shrinkage leads to unrealistically high stress values in the case of linear matrix behaviour. As shown, the polymer behaviour is thermo-mechanically complex, and yielding and degradation occur. After matrix yielding, a redistribution of stress takes place, leading to a more uniform stress distribution.

Also, it was shown that the failure behaviour is largely influenced by the residual stress state itself and should not be neglected. The following conclusions can be made:

- The failure behaviour changes in the case of a superposition with mechanical tensile load. The process-induced stresses change the location of maximum stress.
- During the process, degradation and damage appears in the matrix and the fibre matrix interface. This degrades the resulting composite strength and stiffness.

Further in chapter 6, a simulation strategy was developed using an incremental viscoelastic material model which depends on time temperature polymerisation and fibre volume content. This was chosen in order to promote a computational effective model which is applicable to analyse large problems, as demonstrated by the composite multispar high-lift flap structure. The input parameters of this model are the separate fibre / matrix properties and the fibre volume content. Therefore, it is relatively simple to characterise the necessary parameters, because it is always simpler to characterise the viscoelastic material behaviour of an isotropic material, rather than an anisotropic material. The main advantage of this model is that it provides, not only the results of process-induced distortions and residual stress. There is a full characterisation on the homogenised ply level performed on each time-step of the analysis. Therefore, the model can deliver local changing properties of the part, which can support the structure design in order to be more reliable. The developed model was validated in two test cases. In the first validation test case, the correct representation of the changing engineering properties dependent on the degree of cure has been proven. The following results were found:

- Using an experimental tensile test, an influence of process-induced stresses can be found, related to the variation in curing temperature.
- The developed simulation method can be used to analyse the manufacturing process and provides the following results with reliable accuracy: total degree of cure (<3%), resulting glass transition temperature (<5%), resulting engineering constants (<9%), resulting coefficient of thermal expansion (<25%), process-induced stresses (no reliable experimental data available)

In the second test case, the application of different FE elements, such as composite volume elements or composite shell elements have been discussed and compared with experimental results. The second test case provides following results:

- The performed validation case with the unbalanced stacking was suitable to show the influence of the thermal and chemical shrinkages on process-induced deformations.
- The preferred discretisations are composite volume shell elements.
- The applicability of shell elements is restricted if transverse shear effects are not dominating and deflections are not large. This limits the application to symmetrical laminates or laminates with slightly disturbances of the symmetric conditions.

Following in chapter 7, the developed simulation strategy and material model was used to analyse the manufacturing process of the composite multispar high-lift flap and it were possible to achieve reliable results. One of the main problems during the analysis of the flap was found between the so-called "As-Planned" configuration and the "As-Built" configuration. It was investigated that modelling complex problems like this should always be in close cooperation with the real process, because during the manufacturing, a permanent uncertainty exists and the process is very sensitive to layup or preform errors, temperature changes, etc. Therefore, the model is suitable to analyse the process. The following results were found:

- The simulation strategy and the material model achieve reliable results and can reproduce the physical behaviour of the component.
- The model is suitable to perform sensitivity analysis on risk parameters to achieve a higher process understanding.
- The model is convenient for virtual studies for warpage compensation.
- The highest uncertainty of the model exists between the "As-planned" and "As-built" configuration. This means that, for example, if preform errors are not implemented into the FE model the results will be erroneous. Consequently, the quality management system which should document the process must be highly sophisticated.

In the last chapter different nondeterministic methods have been demonstrated to change the development process situation from an experience-based system to a knowledge-based system. A virtual design of an experiment was used to derive the sensitivity of certain parameters of the material model. The standard deviation of the material parameters was used to show the important parameters in order to improve the accuracy of the developed model. In a second step, Monte Carlo simulation was used to analyse the variability of the process. This is an interesting application, because it is possible to derive required tolerances of process parameters. In the last part of the chapter, a general process optimisation was demonstrated with different objective functions. This final chapter closes with a suggestion of a process development procedure using nondeterministic methods for further applications.

Consequently, the process simulation model is suitable as a more precise starting point for the structural design, because it considers the deformed dimensions, the residual stresses and local variation of material properties, based on the manufacturing process. Therefore, it is also applicable for requalification of parts with failed process parameters. The implementation of the process simulation into the general development process will improve cost efficiency and quality.

10 Abbreviation

RTM	Resin Transfer Moulding
CMF	Composite Multispar Flap
DOE	Design Of Experiment
MCS	Monte Carlo Simulation
LFA	Laser Flash Analysis
DSC	Differential Scanning Calorimetric
TMA	Thermo Mechanical Analysis
DMA	Dynamic Mechanical Analysis
MDSC	Modulated Differential Scanning Calorimetric
CTE	Coefficient of Thermal Expansion
CSC	Chemical Shrinkage Coefficient
DEA	Dielectric Analysis
DOF	Degree Of Freedom
LCF	Liquid Composite Manufacturing
DMS	Strain gauges
TTS	Time Temperature Superposition
WLF	Williams Landel Ferry equation
CLT	Classical Laminate Theory
GCM	Globally Convergent Method
SQP	Sequential Quadratic Programming
ConLin	Convex Linearization
RTI	Resin Transfer Infusion
FE	Finite Element
RVE	Representive Volume Element
DTMA	Dynamic Thermal Mechanical Analysis
TA	TA instruments
HT	High Tenacity
CFRP	Carbon Fibre Reinforced Plastics
voxel	volumetric pixel or Volumetric Picture Element
CHILE	Cure Hardening Instantaneously Linear Elastic model
ERE	Resin modulus
PIOM	Resin Poission ratio
EF1	Fibre modulus
EF2	Fibre modulus transverse
GF12	Fibre shear modulus
POIF	Fibre Poission ratio
PHI	Fibre volume content
SCH	Resin chemical shrinkage

ALF1	Coefficient of thermal expansion in fibre direction
ALF2	Coefficient of thermal expansion transverse
ALM	Coefficient of thermal expansion resin
AURAT	Heat rate
ABRAT	Cooling rate
HAL	Curing time
TEMP2	Curing temperature
TEMP3	Room temperature
W1	Fibre angle ply 1
W2	Fibre angle ply 6
RT	Room temperature
Eng.	Engineering
Num.	Numerical
Eq.	Equation

11 Notation

p	Degree of cure
$\dfrac{dp}{dt}$	Reaction rate
k_i	Arrhenius rate constant
S_i	energy of the reaction, constant of eq. 3.1
A_i	Constant of eq. 3.1
R	gas constant
T	temperature
m	reaction order parameter
n	reaction order parameter
b	Diffusion factor, constant in eq. 3.3
T_g	glass transition temperature
T_{g1}	maximum glass transition temperature
T_{g0}	lower glass transition temperature
T_{C1}^*	Fit parameter, glass transition used by Johnston
T_{Ca1}^*	Fit parameter, glass transition used by Johnston
T_{Cb1}^*	Fit parameter, glass transition used by Johnston
T_{C2}^*	Fit parameter, glass transition used by Johnston
T^*	modified glass transition temperature used by Johnston
a_{tg}	Fit parameter ,glass transition used by Johnston
a_r	Fit parameter ,resin modulus used by Johnston
$f_D(p)$	Diffusion factor
k_D	Rabinowith diffusion factor
T^{**}	modified glass transition temperature
$T_{gel(T)}$	Glass transition temperature at the gel point
T_{end}	Final glass transition temperature
η	lower bound parameter
c	shift factor parameter
ζ	rotation factor
G	shear module resin
γ_i	chemical induced shrinkage coefficients (CSC) composite
γ_m	chemical induced shrinkage coefficients (CSC) resin
V_f	fibre volume ratio
E_m	resin modulus
$E_m{}^0$	Not cured resin modulus
$E_m{}^\infty$	Cured resin modulus
C_{ij}	Stiffness tensor
Δ	increment

ε_i, γ_i	strain
$\Delta\varepsilon_{tot}$	strain increment total
$\Delta\varepsilon_{el}$	strain increment mechanical
$\Delta\varepsilon_{th}$	strain increment thermal
$\Delta\varepsilon_{ch}$	strain increment chemical
α_i	coefficient of thermal expansion (CTE) composite
α_m	coefficient of thermal expansion (CTE) matrix
α_f	coefficient of thermal expansion (CTE) fibre
β	Bilinear coefficient of plasticity
$\sigma_{i(t)}$	tensor stress
$\Delta\sigma_{i(t)}$	stress increment
Δt	time increment
$S_{(t)}$	recursive element
σ_M	Von Mises stress
$\sigma_{Y(T)}$	Temperature dependent yield stress
$\sigma_{F(T)}$	Temperature dependent failure stress
τ_M	characteristic relaxation time matrix
τ_∞	characteristic relaxation time matrix (cured at reference temperature)
ϱ, ρ	Relaxation function
r_i	Relaxation homogenisation factor
a_T	Temperature shift factor
τ_{chr_i}	characteristic relaxation time composite
E_{f1}	young's module in fibre direction (fibre)
E_{f2}	young's module in transverse direction of the fibre (fibre)
G_f	shear module (fibre)
υ_{f12}	Poisson ratio (fibre)
α_{f1}	CTE in fibre direction (fibre)
α_{f2}	CTE in transverse direction (fibre)
$E_{m\infty}$	young's module (matrix)
υ_m	Poisson ratio (matrix)
α_m	CTE (matrix)
α_{eq}	Resulting equivalent CTE of a curved section
α_{th}	Thermal CTE of a curved section
α_{ch}	Chemical Coefficient of a curved section
$\Delta\theta$	Spring-in angle
α_T	CTE tangential in a angle section
α_R	CTE radial in a angle section
ΔT	Temperature gradient
ϕ_T	CSC correction tangential in a angle section
ϕ_R	CSC correction radial in a angle section

E_{ij}	Module of the homogenised ply
ν_{ij}	Poission's ratio of the homogenised ply
G_{ij}	Shear module of the homogenised ply
$\xi_{(t)}$	Reduced time
a_i	Fit parameter heat capacity, eq. 3.12-3.14
b_i	Fit parameter thermal conductivity, eq. 3.15-3.17
d_f	Fibre diameter
R_f	Strength fibre
R_m	Strength matrix
$R_\perp$	Transverse strength composite
$R_\parallel$	Strength in fibre direction - composite
$R_{\parallel\perp}$	Shear strength - composite
$c_{p,m}$	Heat capacity - matrix
$c_{p,c}$	Heat capacity - composite, ply level
$\hat{c}_c$	Heat capacity - composite, laminat level
$k_{c,m}$	Thermal conductivity - matrix
$k_{c,i}$	Thermal conductivity - composite, ply level in material KO system
$k_{X,i}$	Thermal conductivity - composite, ply level in global KO system
$\hat{k}_c$	Thermal conductivity - composite, laminat level
ρ_c	Density - composite, ply level
$\hat{\rho}_c$	Density - composite, laminat level
$\dot{q}$	Exothermal heat flux
H_{tot}	Reaction enthalpy
∇	Nabla operator
K_f	Compression modulus fibre
K_m	Compression modulus matrix
K_l	Bulk modulus
ε_{IFF}	Puck inter fibre effort value
θ_{fp}	Fracture plane angle
p_{XX}^-	Puck rake parameter
θ	Ply angle
r_i	Homogenisation factor relaxation
λ	First Lamé constant

12 References

1. H. Purol, A Stieglitz, P. Woizeschke, A. S. Herrmann, (2010), Beschleunigte Prozesskette für die Herstellung von CFK-Spanten in hoher Stückzahl, Proceedings of the 2010 German Congress on Aeronautics and Astronautics, Hamburg, Germany
2. VDI-2014, Part 1-3 Beuth Verlag 2006
3. M. Kleineberg, N. Liebers, M. Kühn, (2013), Interactive Manufacturing Process Parameter Control, Adaptive, tolerant and efficient composite structures Research Topics in Aerospace 2013, pp. 363-372
4. A. Johnston, (1996), An integrate model of the development of process-induced deformation in autoclave processing of composites structures, PhD thesis The University of British Columbia, Vancouver, Canada
5. L. Kroll, (1992), Zur Auslegung mehrschichtiger anisotroper Faserverbundstrukturen. Dissertation, TU Clausthal, 1992
6. D.W Radford, T.S. Rennick, (2000), Separating Sources of Manufacturing Distortion in Laminated Composites, Journal of Plastics and Composites, vol. 19, no. 8, pp. 621-641
7. A.D. Darrow, (2002), Isolating Components of Processing Induced Warpage in Laminated, Composites, Journal of Composite Materials, vol. 36, pp. 2407-2419
8. B.D Harper, Y.Weitsmann, (1984), Effect of residual stress in polymer matrix composite, The Journal Astronautical Sciences, vol. 32(3), pp. 253-267,
9. A.C. Loos, G.S. Springer, (1983), Curing of epoxy matrix composite, Journal of Composite Materials, vol. 17(2), pp. 135-169,
10. R.J. Stango, S.S.Wang, (1984), Process-induced residual thermal stress in advanced fibre reinforced composite laminates, Journal of Engineering for Industry, vol. 106, pp. 48-54
11. R.H. Nelson, D.S.Cairns, (1989), Prediction of dimensional changes in composite laminates during cure, pp.2379-2410
12. T. Spröwitz, J.Tessmer, T Wille, (2007), Process Simulation in Fiber-Composite Manufacturing – Spring-In, NAFEMS Seminar: Simulating Composite Materials and Structures",November 6 - 7, 2007, Bad Kissingen, Germany
13. G. Fernlund, A. Osooly, A. Poursartip, (2003), Finite element based prediction of process-induced deformation of autoclaved composite structures using 2D process analysis and 3D structural analysis, Composite Structure, 62(2), pp. 223-234
14. T.A. Bogetti, J.W. Gillespie, (1992), Process–induced stress and deformation in thick section thermoset composite laminates, Journal of Composite Materials, vol. 26(5), pp. 620-660
15. G. Fernlund, A. Poursartip, G. Twigg, C. Albert, (2002), Experimental and numerical study of the effect of cure cycle, tool surface, geometry and lay-up on the di-

mensional fidelity of autoclave – processed composite parts, Composites – Part A, Applied Science and Manufacturing, vol. 33(3), pp. 341-351

16. N. Zobeiry, (2006), Viscoelastic constitutive models for evaluation of residual stress in thermoset composites during cure, PhD thesis submitted to the University of British Columbia, Vancouver, Canada

17. Y. Abou Msallem, F. Jacquemin, N. Boyard, A. Poitou, D. Delaunay, S. Chatel, (2010), Material characterization and residual stresses simulation during the manufacturing process of epoxy matrix composites, Composites: Part A, vol. 41, pp. 108-115

18. L.W. Moreland, E.H.Lee, (1960), Thermoviscoelastic analysis of residual stresses

19. S.R. White and H.T. Hahn, (1992), Process Modeling of Composite Materials: Residual Stress Development during Cure. Part I. Model Formulation, Journal of Composite Materials 1992; pp. 26;

20. R.A. Schapery, (1976), Stress analysis of viscoelastic composite materials, Journal of Composite Materials, 1, pp. 228-267

21. J. Svanberg, M.Holmberg, J. Anders, (2004) Prediction of shape distortions Part I. FE-implementation of a path dependent constitutive model, Composites: Part A, vol. 35, pp. 711-721

22. T. Blumenstock, (2003), Analyse der Eigenspannungen während der Aushärtung von Epoxidharzmassen, Institut für Kunststofftechnik Universität Stuttgart, Stuttgart, Germany

23. R.A. Schapery (1974), Viscoelastic behavior and analysis of composite materials, Mechanics of Composite Materials, pp. 85-168

24. S. Wijskamp, (2005), Shape Distorsions in Composite Forming, PhD thesis, University Twente, Netherlands

25. J.M. Svanberg, (2004), Prediction of shape distortions. Part II. Experimental validation
   and analysis of boundary conditions, Composites: Part A, vol. 35, 723–734

26. S.R. White, Y.K.Kim, (1996), Process-induced residual stress analysis of AS4/3501-6 composite material, Mechanics of Composite Materials and Structures, vol. 5(2), pp. 153-186

27. P. Prasatya, (2001), A Viscoelastic Model for Predicting Isotropic Residual Stresses in Thermosetting Materials, PhD thesis, University of Pittsburgh

28. M. Zocher, S. Grooves, D. Allen, (1997), A Three Dimensional Finite Element Formulation for Thermoviscoelastic Media. International Journal for Numerical Methods in Engineering 40,

29. Airbus datasheet, 75-t-2-0601, Werkstoff Handbuch Strukturtechnologie und Versuche, Ausgabe 2/ 1.1.2002

30. HexFlow RTM6, Product Data Sheet, (2007), www.hexcel.com

31. M.R. Kamal, S. Sourour, (1973), Kinetics and thermal characterisation of thermoset cure, Polym. Eng. Sci, pp. 13-59

32. M.E.Ryan, A.Dutta, (1979), Kinetics of epoxy cure: a rapid technique for kinetic parameter estimation, Polymer, vol. 20, pp. 203

33. E. Rabinowitch, (1937), Collision, Co-ordination, Diffusion and Reaction Velocity in Condensed Systems," Transactions of the Faraday Society, vol. 33, pp. 1225-1233.

34. H. Purol, (2011), Entwicklung kontinuierlicher Preformverfahren zur Herstellung gekrümmter CFK-Versteifungsprofile, PhD Thesis, University Bremen FB4, Logos Verlag Berlin, ISBN-10: 383252844X

35. Abschlussbericht / Science Report Lufo IV Fact/Vitech WP 4.1, (2011), Author C.Brauner, Faserinsititut Bremen

36. D.Dykemann, (2008), Minimizing uncertainties in cure modeling in composites, PhD Thesis, University of Columbia

37. Kroll, L.: Berechnung und technische Nutzung von anisotropiebedingten Werkstoff - und Struktureffekten für multifunktionale Leichtbauanwendungen. Habilitationsschrift, TU Dresden, 2005..

38. J.M. Balvers, H.E.N. Bersee, and A. Beukers, (2008), Determination of Cure Dependent Properties for Curing Simulation of Thick-Walled Composites, 49th AIAA/ASME/ASCE/AHS/ASC Structures, Structural Dynamics, and Materials Conference, AIAA 2008-2035

39. P. Karkanas, (1997), Cure modelling and monitoring of epoxy / amine resin systems, PhD Thesis, Cranfield University

40. Aduriz, X.A., et al. (2007), Quantitative control of RTM6 epoxy resin polymerisation by opticalindex determination, Composites Science and Technology , vol. 67, pp. 3196–3201.

41. J. Mijovic, C.Lee, (1989), A comparison of chemoeheological models for thermoset cure, Journal of Applied Polymer Science, vol. 38, pp. 2155

42. P.W.K. Lam, M.R. Piggot, (1989), The durability of controlled matrix shrinkage composites part I: mechanical properties of resin matrices and their composites, Journal of Material Science, vol. 24, pp 4068-4075

43. R.C. Armstrong, B.W. James, et al. (1987), Curing characteristics of a composite matrix, Journal of Material Science, vol. 21, pp. 4289-4295,

44. T. Blumenstock, (2002), Analyse der Eigenspannungen während der Aushärtung von Epoxidharzmassen, Dissertation Universität Stuttgart, Institut für Kunststoffprüfung und Kunststoffkunde

45. M. Holst, (2001), Dissertation: Reaktionsschwindung von Epoxidharz-Systemen. Technischen Universität Darmstadt, Fachbereich Chemie

46. B-C. Chern, Tess J. Moon,J. R. Howell and Wiling Tan, (2002) New Experimental Data for Enthalpy of Reaction and Temperature- and Degree-of-Cure-Dependent, Journal of Composite Materials, vol. 36, no. 17/2002

47. A. A. Skordos,I. K. Partridge, (1999), Monitoring and Heat Transfer Modelling of the Cure of Thermoset Composites Processed by Resin Transfer Moulding, Polymer Composites '99, October 6-8, 1999, Quebec, Canada

48. B.C- Chern, T. Moon, et al. (2002), New Experimental Data for Enthalpy of Reaction and Temperature- and Degree-of-Cure-Dependent Specific Heat and Thermal Conductivity of the Hercules 3501-6 Epoxy System, Journal of Composite Materials, vol. 36, no. 17/2002

49. S.C. Liu, Residual Stress Characterization For Laminated Composites, (1999), PhD Thesis, University of Florida

50. L. Khoun, P. Hubert, (2010), Cure Shrinkage Characterization of an Epoxy Resin System by Two in Situ Measurement Methods, Journal of Polymer Composite, vol. 31, pp. 1600-1610

51. T. Hobbiebrunken, B. Fiedler, M. Hojo, S. Ochiai, K. Schulte, (2005), Microscopic yielding of CF/epoxy composite and the effect on the formation of thermal residual stresses, Composite Science and Technology vol. 65, pp. 1626-1635

52. E. Ruiz, F. Trochu, (2006), Numerical analysis of cure temperature and internal stresses in thin and thick RTM parts, Composites: Part A, vol.36, pp. 806–826

53. M. Wenzel, (2005), Spannungsbildung und Relaxationsverhalten bei der Aushärtung von Epoxidharzen, Dissertation TU Darmstadt, Fachbereich Physik

54. H. Schürmann, (2005), Konstruieren mit Faser-Kunststoff Verbunden, second EditionSpringer Verlage, Berlin

55. M. R. Kulkarni, R. P. Brady, (1997), A model of global thermal conductivity in laminated carbon composite, Composites Science and Technology, vol. 57, pp. 277-285

56. G. Springer, S. Tsai, (1967), Thermal Conductivities of Unidirectional Materials, Journal of Composite Materials, vol. 1 pp.166

57. R. Hill, Z. Hashin, (2011) Samtech user manual 13.1.3

58. G. Meder, (1981), Zur praktischen Berechnung linear-viskoelastischer Probleme, Rheol. Acta 20, 517- 525

59. T. Hahn, (1976), Residual Stresses in Polymer Matrix Composite Laminates, Journal of Composite Materials; vol. 10;

60. A. Puck, (1996), Festigkeitsanalyse von Faser-Matrix-Laminaten Modelle für die Praxis, C. Hanser Verlag, ISBN 3-446-18194-6

61. D.W. Radford, (1993), Cure shrinkage induced warpage in flat uni- axial composites, Journal of Composite Technology and Research, vol. 15 (4), pp 290-296

62. I.M. Daniel, T.-M. Wang, D. Karalekas, and J.T. Gotro, (1990). Journal of Composite. Technologies. Res., vol. 172, pp. 12930

63. Lokost/Probec (Luftfahrtforschungsprogramm IV), (2010) Abschlussbericht

64. A. Osooly, (2008), Development and implementation of robust large deformation and contact mechanics capabilities in process modelling of composite, PhD Thesis, University of British Columbia

65. G.A. Twigg, A. Poursartip, G. Fernlund, (2001), Tool-part interaction in composites processing. Part I: experimental investigation and analytical model, to appear in Composites A.

66. C. Brauner, Y. Radovcic, J-P. Delsemme, P. Jetteur, (2008), Advanced non linear failure analysis of a reinforced composite curved beam with delamination and ply degradation, 2nd Int.Conference on Buckling and Postbuckling Behaviour of Composite Laminated Shell Structures with COCOMAT Workshop, 3-5 September, TU-Braunschweig

67. C. Brauner, (2008), Implementation of the Puck failure criterion in the finite element code SAMCEF for non linear failure analysis, SAMTECH Internal Report

68. A.E.H Love, (1888), On the Small Vibrations and Deformations of Thin Elastic Shells, Philosophical Transactions of the Royal Society, vol. 179, pp. 491ff.

69. G. Kirchhoff, (1850), Über das Gleichgewicht und die Bewegung einer elastischen Scheibe", Journal für reine angewandte Mathematik, vol. 40, pp. 51–58

70. R. D. Mindlin, (1951), Influence of rotatory inertia and shear on flexural motions of isotropic, elastic plates, ASME Journal of Applied Mechanics, vol. 18 pp. 31–38

71. E. Reissner, (1945), The effect of transverse shear deformation on the bending of elastic plates, ASME Journal of Applied Mechanics, vol. 12, pp. A68-77

72. M. Bischoff, (1999), Theorie und Numerik einer dreidimensionalen Schalenformulierung, PhD Thesis, Universität Stuttgart, Institut für Baustatik

73. E. Reissner, (1945), The Effect of Transverse Shear Deformation on the Bending of Elastic Plates, Journal of Applied Mechanics, vol. 12, pp. 69–76.

74. J. N., Reddy, (1999), Theory and analysis of elastic plates, Taylor and Francis, Philadelphia

75. C. Li, K. Potter, M.R. Wisnom, and G. Stringer, (2004), Journal of Composite Technologies., vol. 64, pp. 55

76. Huang, Xiaogang, Gillespie, John W., T. Bogetti, Travis, (2000), Process-induced stress for woven fabric thick section composite structures Composite Structures vol. 49

77. Partridge, Ivana ,Karkanas, Panagiotis, (2000), Cure Modelling and Monitoring of Epoxy/Amine Resin Systems / 1. Cure Kinetics Modeling, Journal of Applied Polymer Science, vol. 77,

78. H. T. Hahn, (1976), Residual Stresses in Polymer Matrix Composite Laminates, Journal of Composite Materials; vol. 10;

79. M. Holst, (2001), Reaktionsschwindung von Epoxidharz Systemen, Dissertation University Darmstadt

80. C Brauner, T. B. Block, A. S. Herrmann, (2011), Meso-level manufacturing process simulation of sandwich structures to analyze viscoelastic-dependent residual stresses, Journal of Composite Materials 0021998311410498

81. A. Bebamzadeh, T. Haukaas, R. Vaziri, A. Poursartip and G. Fernlund, (2010), Application of Response Sensitivity in Composite Processing, Journal of Composite Materials, vol. 44,

82. H.Li, R. Foschi, R. Varazi, G. Fernlund, A. Poursatrip, (2002), Probability-Based Modelling of Composites Manufacturing and its Application to Optimal Process Design, Journal of Composite Materials, vol. 36

83. C.G. Kim, E.J.Jun, (1989), Spring-in deformation of composite laminated bends. Proceedings of the 7th international conference on composite materials (ICCM7) pp. 83-88.

84. R. Jones, (1975), Mechanics of Composite Materials, Mc Graw Hill Bok Company, ISBN 0-07-032790-4

85. K. Rohwer, (1988), Transverse Shear Stiffness of Composite and Sandwich Finite Elements. Proc. Int. Conf. „Spacecraft Structures and Mechanical Testing ESA SP-289, 19–21 Oct. 1988

86. C. Albert and G. Fernlund, (2020), Spring–in and warpage of angled composite laminates. Composites Science and Technology, vol. 62(14), pp. 1895–1912, 2002.

87. A.S.Herrmann, (2012), Vorlesungsscript Mechanik der Faserverbundwerkstoffe I, Universität Bremen

88. ISO 11359-2, (1999), International Standard Plastics- Thermomechanical analysis (TMA), Part2: Determination of coefficient of linear thermal expansion and glass transition temperature, first edition 1999-10-01

Bachelor, Master, Diploma or Student Thesis

S1 T. Frerich, (2012), Master Thesis, University Bremen, Charakterisierung und FEM Implementierung der viskoelastischen Eigenschaften von CFK Laminaten

S2 P. B. Soprano, (2012), Diploma Thesis, University of Santa Catarina, Thermo-chemical simulation of a composite aircraft frame manufactured by RTM process

S3 G. F. Larsen, (2010), Diploma Thesis, University of Santa Catarina, Cure Kinetics and Modelling of Epoxy/Amine Resins

S4 K. Schlindwein, (2011), Diploma Thesis, University of Santa Catarina, Analysis of distortions caused for residual stresses during VARTM process

S5 A. Kuntze, (2011), Bachelor Thesis, University Bremen, Integration von faseroptischen Bragg Gitter Sensoren in RTM-Faserverbundbauteile zum Online-Monitoring des Herstellungsprozesses

S6 T. S. Weber, (2012), Bachelor Thesis, University Bremen, Analysis of processes induced stresses of composite materials by variation of manufacturing condition

S8 J. Schweer, (2012), Bachelor Thesis, University Bremen, Charakterisierung und Analyse des Kriechverhaltens von duromeren Faserverbundwerkstoffen

S9 T.Fischer, (2010), Bachelor Thesis, University Bremen, Online monitoring von Dehnungs- und Aushärtevorgängen während der Herstellung von Faserverbundmaterialien mittels faseroptischer Bragg-Gitter-Sensoren

S10   P. Woizeschke, (2010), Studienarbeit University Bremen, Die Laser-Flash-Analyse zur Bestimmung der Wärmeleitfähigkeit des Epoxidharzsystems RTM6 in Abhängigkeit des Aushärtegrades

S11   S. Dickhut, (2010), Bachelorarbeit Hochschule Bonn-Rhein-Sieg, Untersuchung der Wärmeübertragung innerhalb eines Kohlenstofffaser Preforms senkrecht zur Textillage

13 List of Figures

13. List of figures

Bisher erschienene Bände der Reihe

Science-Report aus dem Faserinstitut Bremen

ISSN 1611-3861

Alle erschienenen Bücher können unter der angegebenen ISBN-Nummer direkt online (http://www.logos-verlag.de) oder per Fax (030 - 42 85 10 92) beim Logos Verlag Berlin bestellt werden.